S0-BVO-722

Coloring Theories

CONTEMPORARY MATHEMATICS

103

Coloring Theories

Steve Fisk

AMERICAN MATHEMATICAL SOCIETY • PROVIDENCE, RHODE ISLAND

1980 *Mathematics Subject Classification* (1985 *Revision*). Primary 05C10, 05C15, 05C25, 05C70.

Library of Congress Cataloging-in-Publication Data

Fisk, Steve, 1946–
Coloring theories/Steve Fisk.
p. cm. — (Contemporary mathematics, ISSN 0271-4132; v. 103)
Includes bibliographical references (p.).
ISBN 0-8218-5109-8 (alk. paper)
1. Map-coloring problem. 2. Group theory. 3. Low-dimensional topology. I. Title. II. Series: Contemporary mathematics (American Mathematical Society); v. 103.
QA612.18.F57 1989
514'.3—dc20 89-27623
CIP

Printed in the United States of America.
This volume was printed directly from author-prepared copy.
The paper used in this book is acid-free and falls within the guidelines established to ensure permanence and durability. ♾

10 9 8 7 6 5 4 3 2 1 94 93 92 91 90 89

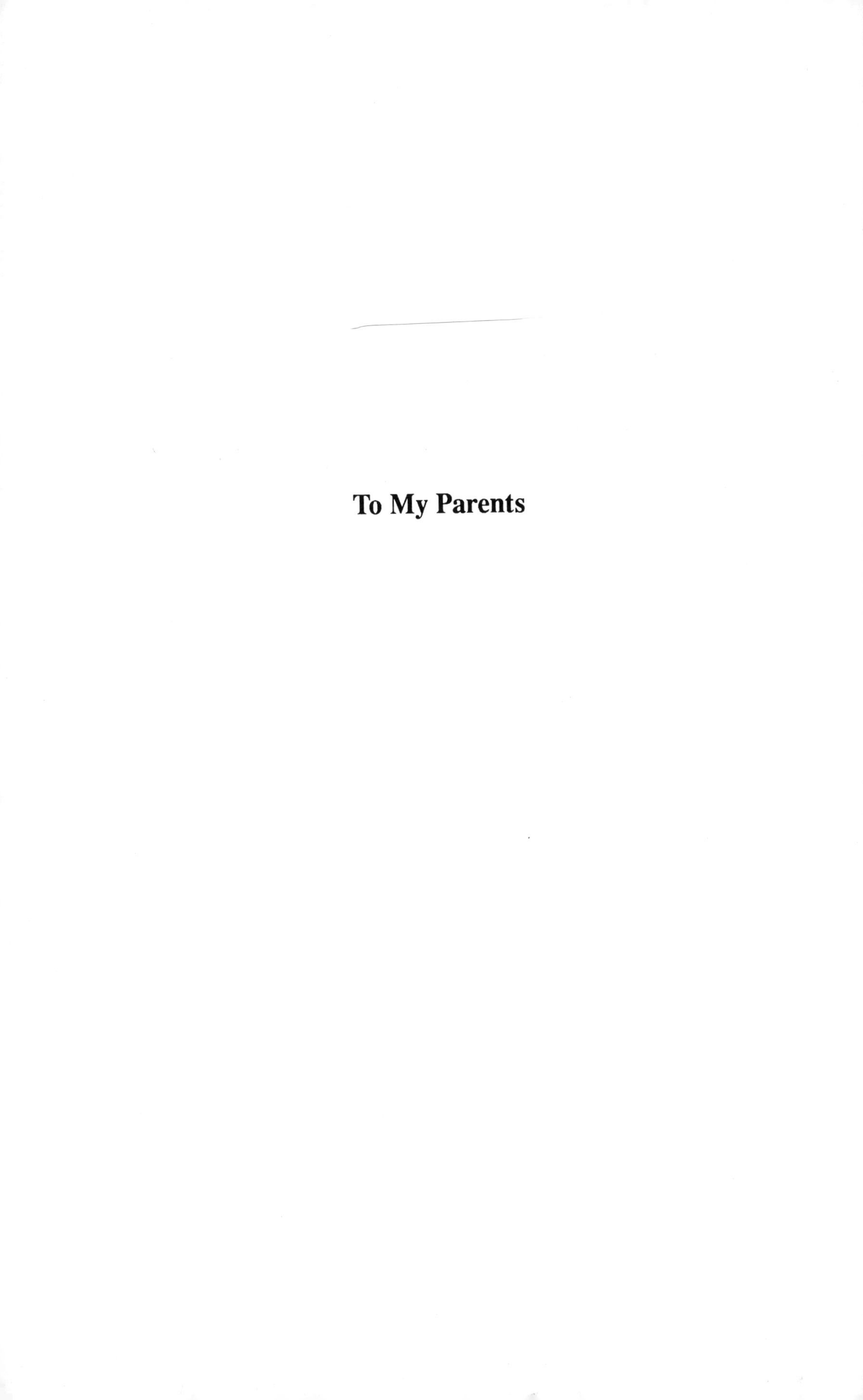

To My Parents

Contents

Preface

The focus of this work is the study of global properties of various kinds of colorings and maps of simplicial complexes. In addition to the usual sorts of coloring, we study colorings determined by groups, colorings based on regular polyhedra, and continuous colorings in finite and infinitely many colors. We are not particularly interested in either the existence of colorings or the number of colorings, but rather in how all the colorings fit together.

A map between two simplicial complexes X and Y is a map from the vertices of X to the vertices of Y such that, for each r, every r-simplex of X is sent to an r-simplex of Y. Every such map is a simplicial map, but not every simplicial map is of this type. An (n+1) coloring of an n-complex is a map from X to the n-simplex Δ^n. This is equivalent to the usual definition. For any n-complex X, we construct an n-complex B(X) whose n-simplices are made from the (n+1)-colorings of X. For any X and Y, the set Hom(X,Y) of maps from X to Y has the structure of an n-complex. In particular, the set of automorphisms of X is another complex Aut(X).

The first chapter determines fundamental properties of B, Hom, Aut, and other constructions, such as cartesian product # , join, hat, etc. For instance, Hom(X,B(Y)) is isomorphic to Hom(Y,B(X)). A particularly important concept is that of reflexivity. We say that X is reflexive if X is isomorphic to B(B(X)). There is a natural map from X to B(B(X)), but it usually is not an isomorphism. A simple result is that if X is reflexive, then Aut(X) and Aut(B(X)) are isomorphic. An interesting calculation is that if R is the 6-regular triangulation of the plane, then Aut(R) is isomorphic to six disjoint copies of R. Next, we introduce composition, and wreath product. B applied to the composition(resp. wreath product) of X and Y is the wreath product(resp. composition) of B(X) and B(Y). A corollary is that the p-Sylow subgroups of the symmetric group are reflexive.

Chapter 2 is devoted to the complex S_n associated to the symmetric group Sym(n). We show that S_n is reflexive, calculate the space of colorings for many complexes determined by the symmetric group, and determine some connectivity properties of the associated complexes. For instance $B(S_n \# \Delta^{n-1})$ is isomorphic to $2 \cdot S_n \# \Delta^{n-1}$. The topological link of the identity in S_n is a simplicial complex formed from derangements; we are able to show that it is reflexive when the dimension is even.

Geometric structures can determine complexes, and in Chapter 3 we study the colorings of certain complexes determined by sets of points in a finite projective space. Every set of points E determines a complex E, and a dual complex E^d. There is a very satisfying interpretation of the space of colorings of such complexes as a geometric dual. For some E the natural map from E to E^d is an isomorphism. We show that this holds for lines in the hyperbolic quartic in projective 3-space PG(3,q), q even.

Chapter 4 studies the line graphs L(G) associated to cartesian products of complete graphs. L(G) is an (n-1)-associated to an n-regular graph. Many line graphs are reflexive, including $L(K_{2n})$, $L(K_n \# K_n)$, and $L(K_{n,n} \# K_{m,m})$. We generalize line graphs, and show that the line complex associated to the multipartite graph $K_{n,n,\ldots,n}$ is reflexive. We introduce group graphs, which are groups G with a simplicial structure such that the mapping $G \# G \rightarrow G$ given by $(a,b) \rightarrow ab^{-1}$ is a map in the category. We determine all 3-regular graphs that are group graphs.

Chapter 5 investigates reflexive and self-dual complexes. The color spectrum is the set of all integers k for which there is a connected complex X satisfying B(X) = kX. If k is 1, the corresponding X are the self-dual complexes. So far, we only know that 1 and 2 are in the color spectrum. We show that various recursive families are reflexive, including a certain family of n-trees, and two classes of triangulations of the 2-sphere. all reflexive bipartite graphs are determined. We conjecture that all the reflexive triangulations of the sphere are known, and we show that there are infinitely many that are not reflexive.

In the sixth chapter we introduce colorings and maps that preserve a group structure. The model on which we base these generalizations is lines in projective space PG(2,q), where each line determines a q-simplex, and where there are maps (projectivities) between any two such simplices. With this new context, we are able to prove that if X is a set of lines in a projective space, then B(X,PG(2,q)) (only the colorings that respect the projective structure) is the disjoint union of X^d and possibly another piece.

If G is a subgroup of the symmetric group on letters 1,2,...n, a G- Latin Square is an n by n Latin Square whose rows and columns are in G, as permutations. We determine the existance of G-Latin Squares for various groups, and are able to determine all such for the 1-dimensional affine group over F_q.

In Chapter 7 we introduce a topology on the space on complexes. For a finite complex, this is the discrete topology. B(X) now inherits a topological structure. Using a Theorem of Lovasz we show that B applied to an infinite product of simplicial complexes is the disjoint union of the colorings of the terms. This is false if we allow all colorings, not just the continuous ones. We show that the integers are reflexive, as well as the cartesian product of the integers with itself. We define topological Latin squares, and show that the 2-sphere does not support any, but that there are some on the 3-sphere. Finally, we detour into analysis by studying the colorings of the cartesian product of the reals with itself. The space is not reflexive, and there are exceptional colorings given by finite compactly supported measures.

In the last chapter, we introduce colorings based on an arbitrary simplicial complex, with the group structure of any group of automorphisms of the complex. We develop a theory of coloring built on regular polyhedra that exactly parallels the usual theory. In particular, many of the results about four colorings of the 2-sphere are true for cubical colorings of quadrangulations of the 2-sphere.

Properties of the Combinatorial Category

This introductory chapter is concerned with the combinatorial properties of pure simplical complexes, and maps between them which never collapse any simplices. In this category, Hom is a functor, and has all the usual properties. There is also the functor B, composed of all the colorings of a complex which satisfies Hom(X,BY) $\cong$ Hom(Y,BX). Similarly, the collection of all automorphisms of a pure simplical complex is functorial. In addition to these basic operations, we also study the join, wreath products, composition products, direct and inverse limits.

1. Hom and Cartesian Product

In this section we establish general properties of the Hom functor. A pure n-complex is an n-complex with the property that every simplex is contained in an n-simplex. Let X and Y be pure n-complexes. A simplicial map f : X $\rightarrow$ Y is *non-degenerate* if f maps all n-simplices onto n-simplices. Equivalently, no edge (1-simplex) is mapped onto a point. Hom(X,Y) has as vertices all non–degenerate maps from X to Y. A set of maps $\{ f_0 \cdots f_n \}$ is defined to be an n-simplex of Hom(X,Y) iff for all x in X, $\{ f_0(x) \cdots f_n(x) \}$ is an n-simplex of Y.

The Cartesian product X#Y of two pure n-complexes X and Y has vertices $\{ (x,y) | x \in X , y \in Y \}$. The simplices are of the form $\{(x,y) | y \in \sigma \}$ for all vertices x of X and simplices σ of Y and $\{(x,y) | x \in \tau \}$ for all vertices y of Y and simplices τ of X. We write these simplices as $x \# \sigma$ and $\tau \# y$. If X and Y are 1-complexes (i.e. graphs) then this is the usual definition of Cartesian product.

The following result gives the basic functorial properties of Hom and Cartesian product.

Theorem 1

(a) The evaluation map X#Hom(X,Y) $\to$ Y is non–degenerate.

(b) The composition map Hom(X,Y)#Hom(Y,Z) $\to$ Hom(X,Z) is non–degenerate.

(c) If f:X $\to$ Y is non–degenerate so are

f_* : Hom(Z,X) $\to$ Hom(Z,Y)

f^* : Hom(Y,Z) $\to$ Hom(X,Z)

(d) Hom(Z#Y,X) $\cong$ Hom(X,Hom(Y,Z))

(e) If f : A $\to$ B and g : C $\to$ D are non–degenerate so is f#g : A#B $\to$ C#D.

Proof

(a) The evaluation map e sends (x,f) to f(x). To check that e is non–degenerate, we must check it on the two types of simplices of X#Hom(X,Y). For a simplex of the form σ#f, e(σ#f) = f(σ), which is a simplex of Y since f is non–degenerate. Consider a simplex of the form x#τ, where $\tau = (f_0, \cdots f_n)$, e(x#τ) = $\{ f_0(x), \cdots f_n(x) \}$. This is a simplex of Hom(X,Y) by definition of the simplicial structure of Hom(X,Y). It was the desire to have the evaluation map non–degenerate that originally led to the definition of Hom.

(b),(c)

These are easy and omitted.

(d) Let f: X#Y $\to$ Z. Define $\tilde{f}$: X $\to$ Hom(Y,Z) by $\tilde{f}$(x)(y) = f(x,y). $\tilde{f}$ is easily seen to be non–degenerate. Conversely, given $\tilde{f}$, define f(x,y) = $\tilde{f}$(x)(y). This is non–degenerate and so we get an isomorphism.

(e) f#g is defined as f#g(a,b) = (f(a),g(b)). This is clearly non–degenerate. □

In order to get further results with Hom, we must make some mild connectivity assumptions. A top simplex of X is a simplex of highest dimension in X. Define X to be

2-path connected if for any top simplices A, B of X there are top simplices A $= T_1, \cdots, T_k =$ B such that for $1 \leqslant i \leqslant k-1$ T_i and T_{i+1} have at least two vertices in common. Note that X#Y is never 2-path connected. $\coprod$ is disjoint union.

Lemma 2

If X is 2-path connected, Hom(X,Y#Z) = Hom(X,Y)#Z $\coprod$ Hom(X ,Z)#Y.

Proof

Let f:X $\to$ Y#Z. If D is a top simplex of X, f(D) is either of the form σ#z or y#τ for σ a top simplex of Y, τ a top simplex of Z, $z \in Z$, $y \in Y$. Suppose that f(D) = σ#z, and let T be a top simplex of X meeting D in at least two vertices. Since f(D) $\cap$ f(T) has at least two points, f(T) can not be of the form $y'\#\tau'$, for $y'\#\tau' \cap \sigma\#z$ has at most one point. Since X is 2-path connected, all top simplices of X map to Y#z.

Next we check that the correspondence is non–degenerate. Let $(f_0, \cdots f_n)$ be a top simplex of the left hand side. We first claim that all the f_i's map to the same term of the right hand side. Suppose $f_1(x) = g(x)\#z$ and $f_2(x) = y\#h(x)$. Since these are adjacent for all x, either g(x) = y or h(x) = z. It follows that for three vertices in a simplex of X, there is a pair a,b for which g(a) = g(b) =y or h(a) = h(b) = z, and so f_1 and f_2 are not adjacent. It follows that the two terms of the right hand side are disjoint.

Now we have that each $f_i(x) = g_i(x)\#z_i$, and all the z_i are equal. The g_i form a simplex of Hom(X,Y), and so the correspondence is non–degenerate. □

Let Δ^n denote the n-simplex. If n is clear from context, we will simply write Δ. The vertices of Δ are labeled 1,2, ... ,n+1.

Lemma 3

If A and B are 2-path connected, then a map $f:A\#B \to C\#D$ factors in one of these ways :

(a) $\gamma\#\delta : A\#B \to C\#D$; $\gamma:A \to C$; $\delta:B \to D$

(b) $\gamma\#\delta : A\#B \to C\#D$; $\gamma:A \to D$; $\delta:B \to C$

(c) $A\#B \to C \to C\#D$

(d) $A\#B \to D \to C\#D$

Proof

By 2-path connectivity, it suffices to assume that $A = B = \Delta$. Assume that $f(1,j) = c\#g(j)$ for $g: \Delta \to D$ and $f(i,1) = h(i)\#d$ for $h : \Delta \to C$. Note that $h(1)=c$ and $g(1)=d$. Either all first or second coordinates of $f(\Delta\#j)$ are equal, so it is either $c\#f_1(j)$ or $w\#g(j)$. Similarly, $f(i\#\Delta)$ is either $h(i)\#u$ or $f_2(i)\#d$. Since $h(1) = c$, $h(i) \neq c$ for $i \neq 1$. Consequently, $f(i,j)$ is either $c\#d$ or $h(i)\#g(j)$. If any $f(i,j)$ is $h(i)\#g(j)$, then they all are and we have case (c). If none are then $c\#d$ occurs twice in the image of some simplex, and the map would not be non–degenerate.

By the above, we may assume that all simplices of $\Delta\#\Delta$ map to simplices of the form $\sigma\#p$. If so, then $f(\Delta\#j) = f_3\#d$ and $f(j\#\Delta) = f_4\#d$ for all i and j. Write $f(i,j) = H(i,j)\#d$. H is a map $\Delta\#\Delta \to C$, and so we have case d. □

Corollary 4

If A and B are 2-path connected, then $\mathrm{Hom}(A\#B,C\#D) = \mathrm{Hom}(A\#B,C)\#D \coprod \mathrm{Hom}(A\#B,D)\#C \coprod \mathrm{Hom}(A,C)\#\mathrm{Hom}(B,D) \coprod \mathrm{Hom}(A,D)\#\mathrm{Hom}(B,C)$

it is worthwhile to note that since A#B is not 2-path connected, we get two more terms than afforded by Lemma 2.

2. The Coloring Functor B

This section introduces the coloring functor B, and establishs some of its properties. B was originally introduced to study the four colorings of a planar triangulation. It has turned out to be a much more general functor and a very interesting one.

If Δ^n is the n-simplex, a coloring of the pure n-complex X is a non−degenerate map f:X $\to \Delta^n$. A vertex of B(X) is a set $f^{-1}(p)$, where f is a coloring of X and $p \in \Delta^n$. The n-simplices correspond to colorings of X. If the vertices of Δ^n are $p_0, \cdots, p_n$, then $\{f^{-1}(p_0), \cdots, f^{-1}(p_n)\}$ is an n-simplex of B(X).

We summarize some of the elementary properties of B [Fisk, 1980]. B is a contravarient functor on the category of pure n-complexes and non−degenerate maps. There is a natural transformation Φ which gives a map $\Phi:$ X $\to B^2(X)$, where $B^2(X) = B(B(X))$. If x is a vertex of X, then Φ(x) is the set of all vertices of B(X) which contain x.

If Φ gives an isomorphism between X and $B^2(X)$, then we say that X is *reflexive.* One of the important problems is the determination of the reflexive complexes. If there is an isomorphism between X and B(X), then we say that X is *self-dual.* As opposed to the reflexive complexes, there are very few self-dual complexes.

A vertex v of B(X) has the property that v meets every top simplex of X in exactly one point. Unfortunately , if we have a set w of vertices of X such that w meets every top simplex of X, w needn't be a vertex of B(X). We call such a w a *potential vertex of* B(X). Consider the 2-complex X of Figure 1. B(X) is empty, for there is no 3-coloring of X, but there is a potential vertex of B(X).

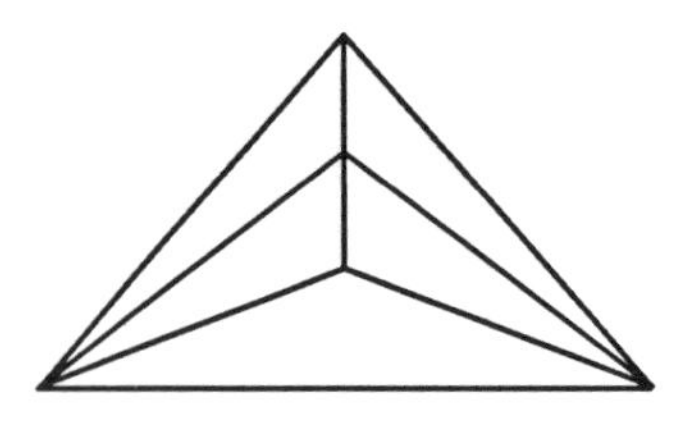

Figure 1

In order to give the next property of B, we introduce a new operation. If X and Y are pure n-complexes, then $X\times Y$ has vertices all pairs (x,y), where $x\in X$ and $y\in Y$. A collection of n+1 pairs $\{(x_0,y_0),\cdots(x_n,y_n)\}$ is an n-simplex of $X\times Y$ iff $\{x_0,\cdots x_n\}$ and $\{y_0,\cdots y_n\}$ are both n-simplices. For all pure n-complexes, we have that

$$B(X\coprod Y)\cong B(X)\times B(Y).$$

If both X and Y are connected ,then we have

$$B(X\times Y)\cong B(X)\coprod B(Y).$$

In Chapter 8 we will generalize this result to products with an infinite number of factors. Our first result relates B and Hom.

Theorem 1

(a) There is a non−degenerate map Hom(X,Y) → Hom(B(Y),B(X)).

(b) Hom(X,B(Y)) ≅ Hom(Y,B(X)) ≅ B(X#Y) Hom(X,Δ^n) ≅ B(X#Δ^n)

(d) If $B^2(X)\cong X$ or $B^2(Y)\cong Y$ then Hom(X,Y) ≅ Hom(B(Y),B(X))

Proof

(a) if f:X → Y is a non−degenerate map , then since B is contravarient there is a map $f*:B(Y)\to B(X)$. If $f_0,\cdots f_n$ is a simplex of Hom(X,Y), we will show that $f*_0,\cdots f*_n$ is a simplex of Hom(B(Y),B(X)). A vertex of B(Y) is a set $g^{-1}(c)$ for g a coloring of X, and c a vertex of Δ^n. By definition, $f*_i(g^{-1}(c)) = f_i^{-1}g^{-1}(c) = (gf_i)^{-1}(c)$.

First, $(gf_i)^{-1}(c)\bigcap(gf_j)^{-1}(c)$ is empty. If x is in the intersection, then both $f_i(x)$ and $f_j(x)$ both lie in $g^{-1}(c)$. Since g is a coloring, no two distinct points of $g(c)$ are adjacent. But since f_i and f_j both lie in the same simplex, $f_i(x)$ is adjacent to $f_j(x)$.

To show that the $(gf_i)^{-1}(c)$ form a simplex of B(X), it suffices to show that every x $\in X$ is contained in one of the $(gf_i)^{-1}$. $\{f_0(x), \cdots f_n(x)\}$ is an n-simplex of Y, and so meets $g^{-1}(c)$. This concludes the proof of (a).

(b) Let f:X#Y $\rightarrow \Delta^n$ and let $c\in\Delta^n$. $f^{-1}(c)$ is a vertex of B(X#Y) and so for each n-simplex D of Y and $x\in X$, $f^{-1}(c)\bigcap x\#D = (x,d)$ for a unique d in D. For fixed x and all possible D's, the set of all these d's gives a vertex of B(Y). Let $\tilde{f}_c : X \rightarrow$ B(Y) be the map which sends x to this vertex.

The association of $f^{-1}(c)$ with $\tilde{f}_c$ is an isomorphism between B(X#Y) and Hom(X,B(Y)). Since $\tilde{f}_c(x)$ and $\tilde{f}_b(x)$ are disjoint if $b \neq c$, $\{\tilde{f}_b(x) | \, b\in\Delta^n \}$ is an n-simplex of B(Y).

We can construct an inverse to this map. Given $g{:}X \rightarrow B(Y)$, define F = $\{(x,y) | \, g(x)\in y\}$. This is a vertex of B(X#Y), and is the desired inverse.

(c) and (d) follow from (b) . □

The map in (a) is generally not an isomorphism. We next have a few results relating Hom, × and ∐ .

Theorem 2

(a) If X is connected, Hom(X × Y,Δ) ≅ Hom(X,Δ) ∐ Hom(Y, Δ).

(b) Hom(U ∐ V,X) = Hom(U,X) × Hom(V,X)

(c) If U is connected, then Hom(U,X ∐ Y) = Hom(U,X) ∐ Hom(V,X)

Proof

Since every map f:X × Y → Δ is of the form f(x,y) = g(x) or f(x,y) = h(y), (a) follows. (b) and (c) are easy and omitted. □

It should be noted that we can not replace Δ^n by an arbitrary complex Z in (a), for there are easy examples where $X\times Y \to Z$ is not a projection from X or Y. For instance, take $Z = X\times Y$.

Our next result shows that coloring cartesian products of 2-complexes is not very interesting. Let B(Z) have $|B(Z)|$ colorings ; equivalently, this is the number of top dimensional simplices of B(Z). Let kX denote k disjoint copies of X.

Theorem 3

If X and Y are connected 2-complexes, then $B(X\#Y) = 2 \cdot |B(X)| \cdot |B(Y)| \cdot \Delta^2$.

Proof

Figure 2 shows the two different three colorings of $\Delta^2\#\Delta^2$. Note that the colorings can be obtained by adding or subtracting the coordinates as given in the middle part of the figure. Given any coloring f of X and any coloring g of Y, we may form the colorings $f+g(x,y) = f(x)+g(y)$ and $f\text{-}g(x,y) = f(x)\text{-}g(y)$. This gives us the correct number of colorings.

To see that there are no more colorings, let S be a triangle of X, T a triangle of Y, s a vertex of S, t a vertex of T, f_y a coloring of s#Y, and f_x a coloring of X#t. A coloring of $\Delta^2\#\Delta^2$ is determined by knowing the coloring on any two triangles. If we know a coloring on S#T, then if S' meets S in p, then since we know the coloring on p#T and S#t, we know the coloring on S'#T as well. By connectivity, if we know the coloring on S#T, f_x , f_y, then we know the coloring.

Since knowing a vertex of $B(\Delta^2\#\Delta^2)$ determines the other two vertices of the triangle containing it, we see that a vertex of B(X#Y) uniquely determines the coloring that it belongs to. Hence, every vertex of B(X#Y) is in a unique triangle, so all the triangles are disjoint. □

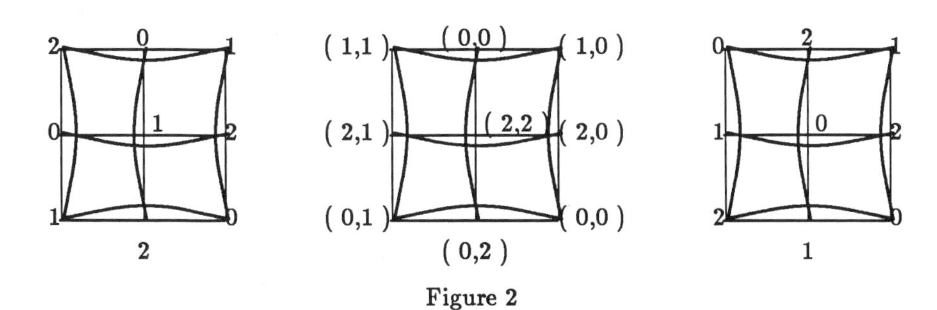

Figure 2

We conclude with an example. Consider the 2-complex of Figure 3. To each of the seven triangles, add a tetrahedron, and denote the new vertex of the tetrahedron added to triangle x,y,z by xyz. The resulting 3-complex X has 7 tetrahedra and 14 vertices. X also has 7 4-colorings, and they are given in the figure. Let the colorings of the first row be a,b,c and d,e,f,g in the second row. Let the vertex of B(X) contained in a coloring α and containing a vertex v of X be denoted αv. For instance, a1 consists of A,C,EFG, and DEG. This is also b1, so the tetrahedron of B(X) corresponding to a intersects the one corresponding to b in at least one vertex. In fact, a3 = b3, so they intersect in exactly two vertices.

The identification of X with B(X) is as follows: A = b2 = c2 ; B = a1 = b1 ; C = a2 = f2 = g2 ; D = e3 = f3 ; E = c1 = d1 = e1 ; F = a3 = b3 = c3 = d3 = g3 ; G = d4 = e4 =f4 = g4. For instance, coloring d determines four vertices d1,d2,d3,d4 of B(X), and these correspond to E,EFG,F,G.

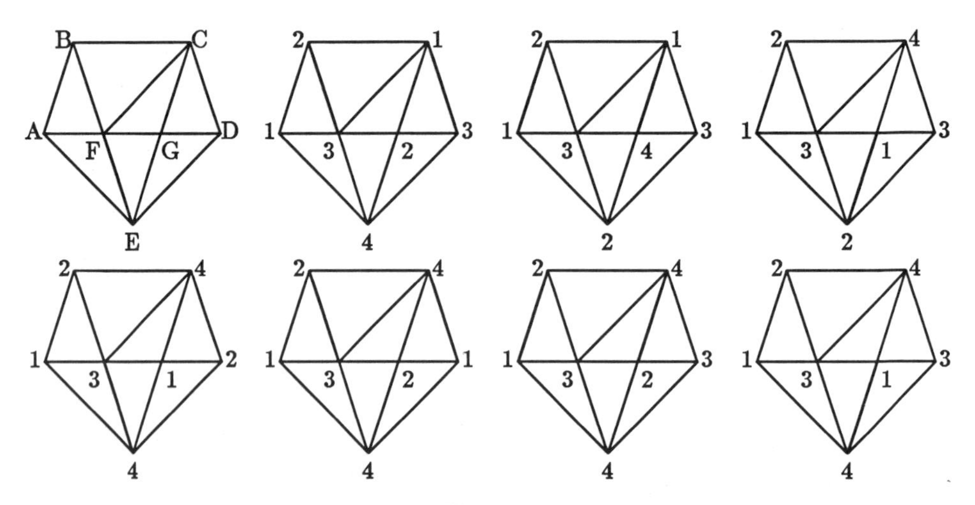

Figure 3

3. The Automorphism Complex

In this section we introduce the complex built out of all automorphisms of a complex X. After deriving some relationships between Aut and B, we give several examples of Aut(X), the most important being S_n, the space of all automorphisms of Δ^n. We also determine all those triangulations X of the 2-sphere for which Aut(X) is non-empty.

If X is an n-complex, define Aut(X) to be the subcomplex of Hom(X,X) with n-simplices the n-simplices of Hom(X,X) whose vertices are all automorphisms. It is possible that there are automorphisms of X which are not vertices of Aut(X), because they do not lie in an n-simplex composed entirely of automorphisms. This happens only if Aut(X) has no n-simplices at all.

Theorem 1

(a) There is a non–degenerate map Aut(X) $\rightarrow$ Aut(B(X))

(b) There is a non–degenerate map Aut(X)#Aut(X) $\rightarrow$ Aut(X) and hence Aut(X) as a group acts on Aut(X) as a complex.

(c) If B^2(X) $\cong$ X then Aut(X) $\cong$ Aut(B(X)).

(d) If A and B are connected, then $\mathrm{Aut}(A \coprod B) = \mathrm{Aut}(A) \times \mathrm{Aut}(B)$ if $A \neq B$ and $4\mathrm{Aut}(A) \times \mathrm{Aut}(A)$ if A=B.

Proof

(a) follows from Theorem 1.1.1b. For (b), define the map to be α(g,h) = gh, where gh is composition. α applied to the n-simplex $\{(g,h_0), \cdots ,(g,h_n)\}$ and evaluated at x in X is $\{gh_0(x), \cdots ,gh_n(x)\}$. By assumption $\{h_0(x),...h_n(x)\}$ is a simplex, and g applied to an n-simplex is an n-simplex, so we've verified that one sort of simplex of Aut(X)#Aut(X) is preserved by α. α applied to the other kind of simplex, evaluated at x, is $\{g_0(h(x)), \cdots g_n(h(x))\}$ and this is also a simplex of X since it is evaluation of $\{g_0, \cdots g_n\}$ at h(x).

(c) follows from Theorem 1.1.2d . (d) is easy and omitted. □

Examples

1. S_n

Let S_n denote $\mathrm{Aut}(\Delta^{n-1})$. The vertices of S_n are the elements of the symmetric group Sym(n) on n letters. If $\sigma_1, \cdots \sigma_n$ is a simplex of S_n, then for any x, all the $\sigma_i(x)$ are distinct, and exhaust the vertices of Δ^{n-1}. If we write the σ_i as permutations, and put them in an n by n matrix as the rows, then we see that an n-simplex of S_n is a Latin square. S_n will be the subject of Chapter 2.

We can describe S_n for small n. S_3 has six vertices, and is given in Figure 1. S_4 has 24 vertices and 24 tetrahedra. It is too complicated to draw directly, but here are two helpful representations. Figure 2 gives the link of the identity. We see every vertex meets 4 tetrahedra, and is adjacent to 9 other vertices. To get a full picture of S_4 ,start with $K_{3,3}$, and replace each vertex by a tetrahedron. Replace each edge of $K_{3,3}$ by Figure 3, where the three edges at a point correspond to three different pairs of edges.

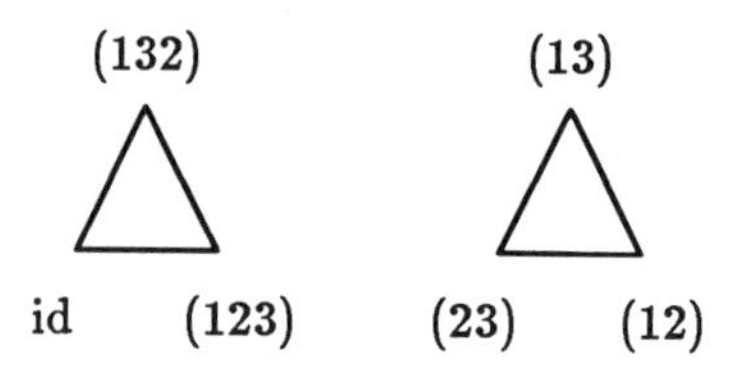

Figure 1

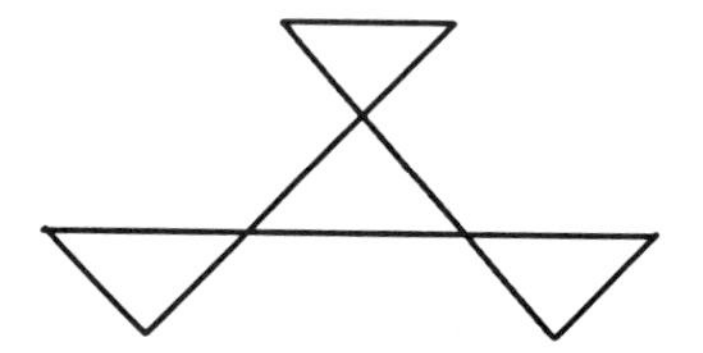

Figure 2

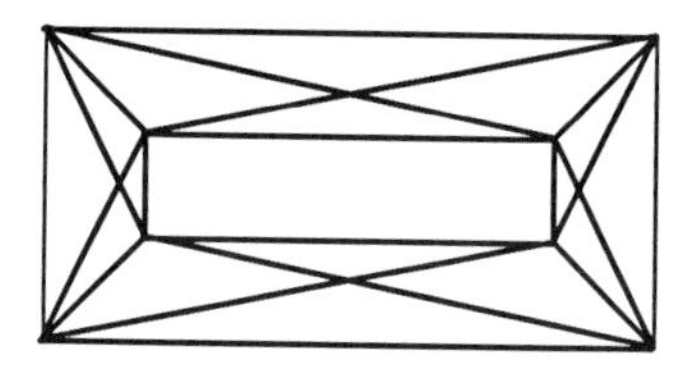

Figure 3

2. The Circle

Let P_n denote the cycle with n vertices. From [Fisk, 1982], $\mathrm{Aut}(P_n)$ is either $2P_n$ if n is not 4, or $4\Delta^0 * 4\Delta^0$ if n is 4.

3. The 2-Sphere

If f and g are automorphisms of X, and lie in the same n-simplex, then for all x, we have f(x) and g(x) are adjacent. If we have two automorphisms f and g of X such that for all x in X, we have f(x) is adjacent to g(x), we say that f and g are adjacent. It is possible that two automorphisms are adjacent, but do not belong to the same top simplex of Aut(X).

Theorem 2

The only triangulations of the 2-sphere which admit two adjacent automorphisms are

(1) the triangle

(2) the tetrahedron

(3) stacked octehedra (see proof for definition)

Proof

Let the triangulation be T. If an automorphism g is adjacent to h, then the identity is adjacent to $g^{-1}h$, so it suffices to take one of the automorphisms as the identity. σ is adjacent to the identity iff for all x, x is adjacent to $\sigma(x)$.

Suppose that there is an edge ab such that $\sigma(a) = b$ and $\sigma(b) = a$. If ab lies in triangles abc and abd, then since ab is sent to itself, either $\sigma(c) = c$ *or* $\sigma(c) = d$. It can not be the former, for σ is adjacent to the identity. Consequently $\sigma(c) = d$ and $\sigma(d) = c$. It follows that $\sigma^2(a) = a$, $\sigma^2(b) = b$ and $\sigma^2(c) = c$. Since any automorphism of a triangulation of the 2-sphere is determined by its action on one triangle, σ^2 is the identity.

The idea behind the proof is to study the orbits determined by σ. If x is any vertex of T, and $\sigma^n(x) = x$, then $x, \sigma(x), \sigma^2(x) \cdots, \sigma^{n-1}(x)$ is a closed path in T. The path might not be a cycle, for it might cover itself several times. If k = ord(x) is the smallest k such that $\sigma^k(x) = x$, then k is the length of the cycle. The above paragraph shows that if ord(x) = 2 for some x then ord(x) is 2 for all x in T.

Suppose that there is an x such that ord(x) is greater than 2. The above cycle splits T into two disjoint pieces A and B. Since there are no edges from the interior of A to the interior of B, if A has any interior vertices, then $\sigma(A) = A$ and $\sigma(B) = B$. If y is an interior vertex of A, then we get a cycle $y, \sigma(y) \cdots$ which is carried into itself by σ, and so lies entirely in the interior of A. Continuing, we get a cycle z, $\sigma(z)$ $\cdots$ which divides the sphere into two pieces C and D, where C has no interior vertices, and $\sigma(C) = C$. C is therefore a triangulated polygon, and so has a vertex of degree 2. If C is not a triangle, there is a vertex of degree greater than 2. However, any point of C is carried to any other point of C by some power of σ, so C is a triangle.

We now see, under the assumptions of the above paragraph, that $\sigma^3 = 1$. Consequently, all cycles are triangles, and we can construct T as follows. Take an octehedron, deform it into a triangular prism, and stack some number of these prisms on top of one another. This triangulation is the stacked octehedron of part (c).

Now suppose that ord(x) is greater than 2, x determines a cycle Z, and Z splits T into A and B, where A and B have no interior points. If $\sigma(A) = A$ then the previous argument shows that T is a triangle. If $\sigma(A) = B$, then $\sigma^2(A) = A$. Every other vertex of Z in A must have degree 2. The remaining vertices all have the same degree. It is clear that Z is either of length 4 or 6, and T is either the tetrahedron or the octahedron.

We are left with the case where $\sigma^2 = 1$. Let two adjacent triangles be xyz and xyw. Since $\sigma(z)$ = w, w and z are joined by an edge. The graph formed by x,y,z and w is the complete graph on 4 vertices , and so determines four regions. σ carries each region to a different one, so no region has any interior vertices. Consequently, T is the tetrahedron.

□

If T is a triangulation of an arbitrary surface, we conjecture that except for the torus, Klein surface and sphere, there are only a finite number of T such that Aut(T) is non-empty. The six-regular triangulations on the torus and Klein surface have automorphisms adjacent to the identity.

4. Cubic Graphs

A cubic graph is a graph with all vertices of degree 3.

Theorem 2

There are only two classes of cubic graphs G with Aut(G) non-empty.

Proof

Let σ be adjacent to the identity, and let a cycle C be p , $\sigma(p)$ $\cdots$ $\sigma^r(p)$ where r is greater than 1. Let q be the unique point adjacent to p, distinct from $\sigma(p)$ and $\sigma^{-1}(p)$ Suppose that q is not in C. Then q determines a cycle D, and the points $\sigma^i(p)$ of C are adjacent to $\sigma^i(q)$ of D. Thus G is two concentric circles joined by spokes.

Suppose that q is in C. If q is $\sigma^i(p)$ for some i, then we can describe G as follows : take a circle of length r+1 , and joint every point to the point i away along the circle in the clockwise direction.

Suppose that all cycles have length 2. G is easily seen to be either the first case, or the second where the length is even, and i is one half the length. Examples of these classes are shown in Figure 4.5.1. □

5. The 6-regular triangulation of the plane.

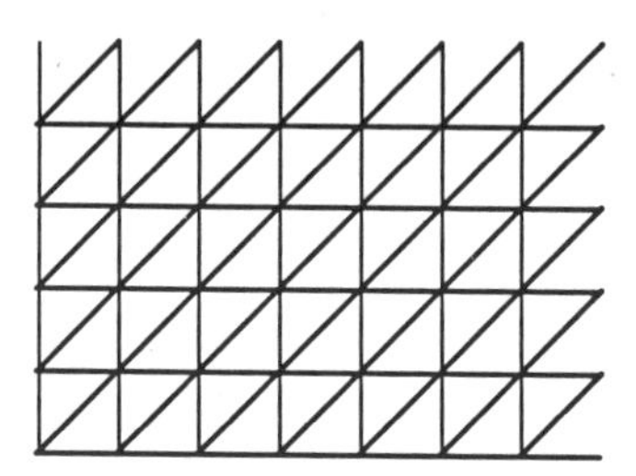

Figure 4

R(6) is the triangulation of the plane by equilateral triangles. Every vertex has degree 6. Its convenient to choose a different coordinate system so that the vertices are the lattice points of the plane and the vertex (a,b) is adjacent to (a+1,b), (a+1,b+1), (a,b+1), (a-1,b), (a-1,b-1), (a,b-1). A portion of R(6) is shown in Figure 4. An automorphism is determined by where the triangle (0,0) , (1,0) , (1,1) is sent. Assume the automorphism σ is adjacent to the identity, and sends this reference triangle to x,y,z. For the moment, assume that x is (1,1). If $\sigma(1,1) = (0,0)$, then (1,0) must be either sent to itself or (0,1). These are both impossible, so $\sigma(1,1) \neq (0,0)$. Similarly, $\sigma(1,0) \neq (0,0)$. Since the edge $\sigma(1,1)\,\sigma(1,0)$ is adjacent to (1,1)(1,0), it follows that there are only two choices: $\sigma(1,1) = (2,2)$, $\sigma(1,0) = (2,1)$ and $\sigma(1,1) = (2,1)$, $\sigma(1,0) = (2,0)$. The latter one is not possible for (1,0) and (0,0) have images that are not adjacent. Consequently, the mapping can be given as $\sigma(a,b) = (a+1,b+1)$. Denote the automorphism $\sigma(a,b) = (a+c,b+d)$ by A(c,d).

We see that there are six automorphisms adjacent to the identity: A(1,0), A(1,1), A(0,1), A(-1,0), A(-1,-1), A(0,-1). Furthermore, each of these is adjacent to the preceding one, and A(0,-1) is adjacent to A(1,0). To see this, note that A(0,0) is adjacent to A(1,1) and if we apply A(-1,0) to each of these we get A(-1,0) adjacent to A(0,1). Thus every vertex of Aut(R(6)) has degree six, and locally looks like R(6). Since R(6) is simply connected, it follows that Aut(R(6)) is the disjoint union of some number of copies of R(6). In any one of the copies, there is exactly one of the triangles which sends the reference triangle to itself , so there are six copies of R(6).

Theorem 3

$$\mathrm{Aut}(R(6)) \cong 6R(6)$$

4. Hat and Join

In this section we introduce the hat and join operations, and derive a few of their properties. If X is a complex, V(X) is the set of vertices of X. If X is a pure n-complex, and Y is a pure m-complex, then X*Y has vertex set $V(X) \cup V(Y)$, and the vertex set of a simplex is the union of the vertex sets of a simplex of X and a simplex of Y. Join behaves nicely under coloring :

$$B(X*Y) \cong B(X)*B(Y)$$

If X is a pure n-complex, then $\hat{X}$ contains X, and has an additional vertex for each n-simplex of X. This vertex is joined to the corresponding n-simplex to form an n+1-simplex. For instance, if P is a polygon with k vertices, then $\hat{P}$ has k triangles, arranged in a circle. The Hat operation arises in a natural way. For instance, if we find all the 3 -colorings of a circle P with k vertices, and then find all colorings of the resulting complex, we get $\hat{P}$, not P.

Theorem 1

Suppose that X is an n-complex satisfying

(1) $B^2(X) \cong X$

(2) For any vertex $w \in X$, if I is an independent set of X which meets every vertex of B(X) containing w, then $w \in I$

then $B^2(\hat{X}) \cong \hat{X}$.

Proof

Suppose that $f : X \to \Delta^n$ is a coloring of X. We get a coloring $\hat{f}$ of $\hat{X}$ by setting $\hat{f}| X = f$ and $\hat{f}(\hat{X}-X) = \Delta^{n+1}-\Delta^n$. $\hat{X}-X = \Lambda$ is the set of all the vertices added by the hat construction. We consider Λ to be a vertex of $B(\hat{X})$.

Suppose that $\hat{F}$ is a coloring of $B(\hat{X})$. The above construction embeds B(X) in $B(\hat{X})$ as $B(X) * \Lambda$, and so $\hat{F}$ induces a coloring F on B(X). Since X is reflexive, F is induced by a simplex of X. A vertex of F consists of all the vertices of B(X) containing some fixed vertex of X. Hence, a vertex of $\hat{F}$ contains all the vertices of B(X) containing some fixed vertex w of X.

Let β be a vertex of $\hat{F}$, and assume that $\Lambda \notin \beta$. β is a collection of vertices of $B(\hat{X})$, exactly one in each n+1-simplex of $B(\hat{X})$. Let $\alpha \in \beta$ and let $\{g_0, g_1, \cdots g_n\}$ be the vertices of any coloring of X, with $g_0 \in \beta$. We construct a new n+2 coloring of X by using the vertices $\{g_0 - \alpha, \cdots, g_n - \alpha, \alpha\}$. This coloring of X extends uniquely to a coloring of $\hat{X}$. If $\alpha \cap g_0 = \emptyset$, then we get a coloring of X with the two members α and g_0 of β in it. Consequently, $\alpha \cap g_0 \neq \emptyset$ for all $\alpha \in \beta$. Taking I = α, the second hypothesis gives that w$\in\alpha$. β therefore consists of all vertices of B($\hat{X}$) containing w.

We thus see that all vertices of F which contain a vertex of B(X) are in the image of Φ. It is easy to check that all these vertices lie in a simplex of X, and that the last vertex consists of all the vertices of B($\hat{X}$) containing Λ. □

Corollary 2

If X is $\Delta^n \times \Delta^n$ or $\Delta^n \# \Delta^n$ then $\hat{X}$ is reflexive.

As an example, it is easy to calculate that $B(k \bullet \Delta^1) \cong 2^{k-1} \bullet \Delta^1$. If we choose k equal to 2, we see $2 \bullet \Delta^1$ is self-dual. By taking joins with Δ^n, we get examples of self dual complexes in all dimensions.

5. Wreath Products and Graph Composition

We now introduce the wreath product $C \int B(D)$ and the composition product C $\circ$ D. Under certain restrictions on D, $B(C \int B(D))$ is the composition $D \circ B(C)$. It is always true that $B(X \circ Y) \cong B(Y) \int B(X)$. Together, these two results imply that many wreath products are reflexive, including the p-Sylow subgroups of S_n.

To motivate the definition, consider the group Sym(h) $\int$ Sym(k). Its elements are $(e_1, e_2, \ldots e_k, \pi)$ where $e_i \in$Sym(h) and $\pi \in$Sym(k). We consider π as a permutation on the k blocks [1...h],...[1...h] and apply e_i to the π(i)-th block. If this permutation is adjacent in S_{hk} to the one determined by $(d_1, d_2, \ldots d_k, \tau)$ then no two elements are sent to the same element. This implies that if $\pi(i) = \tau(i)$ then e_i is adjacent to d_i. We take this as our first definition:

Let H be an (h-1)-complex. The elements of $H\int S_k$ are all ordered sets $(e_1,e_2,\ldots e_k,\pi)$ where $e_i\in H$ and $\pi\in S_k$. $(e_1,e_2,\ldots e_k,\pi)$ is adjacent to $(d_1,d_2,\ldots d_k,\tau)$ iff for each i such that $\pi(i) = \tau(i)$, e_i is adjacent to d_i.

We will generalize this definition, but we must first introduce composition. This is the same operation as graph composition, except that we apply it to simplicial complexes. The composition of two simplicial complexes H and K has vertices all pairs (x,y), with $x\in H$ and $y\in K$. If $\{t_1,\ldots t_k\}$ is a top simplex of K, and $\{s_{i1},\ldots s_{ik}\}$ is a top simplex of K for i = 1...h, then $\{(t_i,s_{ir})|\ 1\leqslant i\leqslant h,1\leqslant r\leqslant k\}$ is a top simplex of $H \circ K$. Less rigorously, we relace each vertex of H by a copy of K and join together two copies of K iff their corresponding vertices are adjacent. The top simplices are obtained by replacing all the vertices of a top simplex of H by a top simplex of K. $H \circ K$ is an hk-1 complex. We can easily calculate $B(H \circ K)$. Let K_p be the copy of K replacing the vertex p of H. If f is a coloring with hk colors of $H \circ K$, a vertex $f^{-1}(c)$ of $B(H \circ K)$ has one vertex in each simplex of $H \circ K$. In every top simplex of H there is a unique vertex p such that $f^{-1}(c)\cap K_p \neq \emptyset$. The set of these p's gives us a vertex α of B(H). For each p, $f^{-1}(c)\cap K_p$ is a vertex of $B(K_p)$. Consequently, a vertex of $B(H \circ K)$ is given by a pair (e,α) where $\alpha\in B(H)$ and e : $|\alpha| \rightarrow B(K)$ is a map from the underlying point set of α to B(K). A collection (e_i,α_i),$1\leqslant i\leqslant hk$ forms a top simplex iff for every $p\in H$ m there are exactly k α_i's containing p, and the k corresponding $e_i(p)$ from a simplex of B(K). This leads to the definition :

$X\int B(Y)$ has vertices all pairs (e,α) with $\alpha\in B(Y)$ and e : $|\alpha| \rightarrow$ X. $\{(e_i\ ,\ \alpha_i),1\leqslant i\leqslant hk\}$ is a top simplex iff every vertex p of Y occurs in exactly h α_i's and the corresponding $e_i(p)$'s form a simplex of X.

As an immediate consequence we have

Theorem 1

$$B(H \circ K) \cong B(K) \int B(H)$$

We check that the two definitions for $H \int S_k$ and $H \int B(\Delta^{k-1} \# \Delta^{k-1})$ agree. The vertices of $H \int B(\Delta^{k-1} \# \Delta^{k-1})$ are of the form (e,α) for $\alpha \in B(\Delta^{k-1} \# \Delta^{k-1})$. There is a unique permutation π such that $\alpha = \{(i,\pi(i)) \mid 1 \leqslant i \leqslant k\}$. If we define $e_i = e(i,\pi(i))$ then (e,α) corresponds to $(e_1, \cdots e_k, \pi)$.

In the definition of $H \int S_k$, we only defined adjacency, not the top simplices. This was because every edge of $H \int S_k$ is contained in a top simplex. Adjacency in the second definition implies adjacency in the first, so the two definitions agree.

Theorem 2

If Φ:K $\rightarrow$ B^2(K) is an isomorphism onto its image, then $B(H \int B(K)) \cong K \circ B(H)$.

Proof

The idea of the proof is to find colorings of subcomplexes and piece them together. Consider the subcomplex of $H \int B(K)$ whose vertices are $<\alpha> = \{(e,\alpha)\}$. If (e,$\alpha$) and (d,$\alpha$) are in the same simplex of $H \int B(K)$, then for each p in α, e(p) is adjacent to d(p). If α has $\#\alpha$ members, we see $<\alpha>$ is isomorphic to $H \times H \cdots \times H$ ($\#\alpha$ times). Let $\alpha_1, \cdots \alpha_k$ be a simplex of B(K). Every (e,α_i) is joined to (d,α_j) as long as i $\neq$ j. Thus, $H \int B(K)$ contains the repeated join $<\alpha_1>*<\alpha_2> \cdots *<\alpha_k>$.

Let F be a coloring of $H \int B(K)$. Considering F restricted to the above join, we see that F uses h colors on each of the $<\alpha_i>$'s, and restricts to a coloring on each $<\alpha_i>$. Every coloring g of $H \times H \cdots \times H$ is given by $g(h_1, \cdots h_t) = \tilde{g}(h_i)$ for some fixed index i and coloring $\tilde{g}$ of H. Consequently, for a given α, there is an index g(α) which lies in α and F(e,α) = f_α(e(g(α))) for f_α a coloring of H and g(α)$\in\alpha$. We use this representation to determine the vertices of B($H \int B(K)$).

Suppose that $F(e,\alpha) = F(d,\beta)$. We have $f_\alpha(e(g(\alpha))) = f_\beta(d(g(\beta)))$. $F(e,\alpha)$ depends only on $e(g(\alpha))$ and on no other point of α other than $g(\alpha)$. If $g(\alpha) \neq g(\beta)$ then we can find e' and d' such that $e'(g(\alpha)) = e(g(\alpha))$, $d'(g(\beta)) = d(g(\beta))$, and for all p where possible, e'(p) is adjacent to d'(p). Consequently, we have that $F(e',\alpha) = F(d',\beta)$ and (e',α) is adjacent to to (d',β). This is a contradiction, and so we have $g(\alpha) = g(\beta)$.

Let $A = \{\alpha \mid \text{there is an } e \text{ such that } f(e,\alpha) = c\}$ for c fixed. There is exactly one member of A in every top simplex of B(K), so $A \in B^2(K)$. If $F(e,\alpha) = F(d,\beta) = c$, then the coloring F is of the form $g_\gamma(e(p))$ for $\gamma = \alpha$, or β and $p \in \alpha \cap \beta$. p thus lies in every member of A, so $A \subset \Phi(p)$. Since no vertex of B(K) is a subset of another, $A = \Phi(p)$.

We now know that if $F(e,\alpha) = c$ then there is a $p \in \alpha$ such that $g_\alpha(e(p)) = c$, and p depends only on c. The set of e's satisfying this is a product of H's for each $q \neq p$, and a vertex of B(H) for p. Call this vertex $r(\alpha)$. Consider a β such that $F(e,\beta) = c$ with corresponding vertex $r(\beta)$. If $r(\alpha) \neq r(\beta)$, some top simplex of H has distinct x and y such that $x \in r(\alpha)$ and $y \in r(\beta)$. We choose e and d such that $e(p) = x$, $d(p) = y$, and for all other relevant q, e(q) is adjacent to d(q). Thus $F(e,\alpha) = F(d,\beta)$ since $e(p) \in r(\alpha)$ and $d(p) \in r(\beta)$, but (e, α) is adjacent to (d,β). Consequently, $r(\alpha)$ is a constant $r \in B(H)$.

We have that a vertex of F, $F^{-1}(c)$, consists of all (e,α) such that $p \in \alpha$, and $e(p) \in r$. We write $F^{-1}(c) = <p,r>$ with $p \in K$ and $r \in B(H)$. Suppose $F^{-1}(c') = <p,s>$. If $x \in r \cap s$, then we can find (e, α) such that $p \in \alpha$, $e(p) = x$. $F(e,\alpha) = c$ since $e(p) \in r$ and also $F(e,\alpha) = c'$ since $e(p) \in s$. Consequently, if $<p,r>$ and $<p,s>$ are vertices of F, then $r \cap s = \emptyset$.

Suppose we have vertices $<p,r>$ and $<q,s>$ of F with p not adjacent to q. By hypothesis Φ is one to one, so there is a vertex α of B(K) such that p and q lie in α. $F(e,\alpha) = c$ iff $p \in \alpha$ and $e(p) \in r$. $F(e,\alpha) = c'$ iff $q \in \alpha$ and $e(q) \in s$. Since $p \neq q$, we may find an e such that (e,α) is colored both c and c'. Consequently, p is adjacent to q.

If $<p,r> = <q,s>$, then $\Phi(p) = \Phi(q)$. From our assumptions, $p = q$. It then follows that $r = s$, so all the $<p,q>$ are distinct.

We can now conclude that for all vertices <p,r> and <q,s> of $F^{-1}(c)$, p is adjacent to q. K has no more than k mutually adjacent vertices, so there are at most k distinct p's occurring in <p,r> = $F^{-1}(c)$. For fixed p, all the second coordinates are disjoint, and hence there are at most h of them Since there are hk colors, we conclude that there is a set of k mutually adjacent vertices of K, and for each of these vertices of K, there is a set of h disjoint vertices of B(H) which form a top simplex of B(H). Since Φ is an isomorphism onto its image, every set of k mutually adjacent vertices of K is a top simplex of K.

So, $B(H\int B(K))$ consists of all pairs <p,r>, with p $\in$ K, r $\in$ B(H). The top simplices are formed by choosing a top simplex of K, and for each point of this simplex, a top simplex of B(H) is chosen. This is precisely K $\circ$ B(H) . $\square$

If one merely assumes that Φ is one to one, then one can show that $B(H\int B(K))$ = W $\circ$ B(H), where W has the same vertices as K, and any collection of k mutually adjacent vertices of K is a top simplex of W.

Corollary 3

If $\Phi : X \to B^2(X)$ is an isomorphism, then

$$B^2(X \circ Y) \cong X \circ B^2(Y)$$

$$B^2(X \int B(X)) \cong B^2(Y) \int B(X)$$

If in addition Y is reflexive, so are X $\circ$ Y and Y $\int$ B(X).

We now apply these results to a special case. Consider a simplex C with p vertices. B(C) = C, and so C $\int$ C is reflexive. Continuing, C $\int$ C $\int$ ...C is reflexive. Choosing the correct number of factors, this has group structure isomorphic to the p-Sylow subgroup of Sym(n). Consequently, the simplicial complexes corresponding to the p-Sylow subgroups of Sym(n) are reflexive.

As a last example, consider the n-dimensional octahedron. It may be realized as

$(S^0)^{*n}$, where S^0 is two disjoint points, and W^{*n} is $W * W \cdots * W$ (n factors). It is not hard to see that $\mathrm{Aut}((S^0)^{*n}) \cong S_n[2^n \bullet \Delta^1]$. By the corollary, $\mathrm{Aut}((S^0)^{*n})$ is reflexive.

6. Limits

In this section we define direct and inverse limits in the category of simplicial complexes and non-degenerate maps. We give examples of these limits, and find some relationships with the coloring functor.

We first define a directed system. Let $\{X_i\}$ be a family of simplicial complexes, where the indices are in some poset. If i $\leqslant$ j, then there is a map f_{ij} from X_i to X_j. These maps satisfy the compatibility condition : if $i \leqslant j$ and $j \leqslant k$ then $f_{ik} = f_{jk} \circ f_{ij}$. f_{ii} is the identity map. We denote this directed system by $\{X,f\}$.

Suppose that we have two directed systems $\{X,f\}$ and $\{Y,g\}$. A map from $\{X,f\}$ to $\{Y,g\}$ begins with an order preserving map h from the poset of the X's to the poset of the Y's. Next, there are maps H_i from X_i to $Y_{h(i)}$ such that for all i $\leqslant$ j we have $H_j \circ f_{ij} = g_{h(i)h(j)} \circ H_i$. One important case is when one of the posets consists of a single element.

Let B(X,f) be the directed system whose members are $B(X_i)$. Since B is a contravarient functor, we take the poset to be the opposite of the poset of the X's. Thus if i $\geqslant$ j for the X's, then there is a map f_{ji} from X_j to X_i. Applying B, we get Bf_{ji} from $B(X_i)$ to $B(X_j)$. BH gives a map from B(Y,g) to B(X,f).

The direct limit of a system is the universal object mapping to the system. The inverse limit is the universal object mapped onto by the system. In more detail ,the direct limit $\lim\limits_{\rightarrow}(X,f)$ is a complex Y such that for any complex Z which maps to $\{X,f\}$, there is a unique map from Z to Y which is compatible with the maps f_i. The inverse limit $\lim\limits_{\leftarrow}(X,f)$ is the complex Y such that if Z is any complex which is mapped to from $\{X,f\}$ then there is a unique map from Y to Z compatible with the maps. If H is a map between two directed systems with the same poset and H is an isomorphism on all terms, then H is an isomorphism on the limits.

Let $\{X,f\}$ be a directed system. The vertices of $\varinjlim(X,f)$ are all collections $\{x_i\}$, one x_i from each X_i such that the x_i are compatible. This means that if $i \leqslant j$, then $f_{ij}(x_i) = x_j$. A set of vertices forms a simplex of the direct limit iff the vertices in each of the coordinates forms a simplex. This definitions of the vertices is the same as the category of sets; what is new is the observation that the definition of simplex makes sense.

In the category of sets, the inverse limit of the X's consists of all collections (x_i), where the x_i's are are maximal collection of members of X which are compatible. Unfortunately, $\varprojlim(X,f)$ is not always the same dimension as the X_i's. Here is a simple example. Let the complexes of the directed system be X,Y,Z, with maps f from X to Y and g from X to Z. X is $2\bullet\Delta^1$ and $Y = Z = \Delta^1$. Let the vertices of X be a,b,c,d with edges ab, and cd. Let Y have vertices 1,2 and Z have vertices 3,4. Define f(a)=1, f(b)=2, f(c)=1, f(d)=2, g(a)=3 ,g(b)=4, g(c)=4, g(d)=3. Since f(a)=f(c)=1, a,c and 1 map to the same point of $\varprojlim(X,f)$. Similarly, the triples b,d,2 ; a,d,3 ; b,c,4 all map to single points of $\varprojlim(X,f)$. Consequently, $\varprojlim(X,f)$ consists of a single point.

If $\{X,f\}$ has a map to any complex Y, then $\varprojlim(X,f)$ exists. As sets we have the maps $\{X,f\}\to\varprojlim(X,f)\to Y$. There is no degeneracy, else the original map is degenerate. If $X = \varinjlim(X,f)$ and some X_i has a coloring, then $B(X) \neq \emptyset$. We get a coloring of X by composing the projection from X to X_i with the coloring of X_i.

Examples

(1) Let X be a pure n-complex, and let X_i consist of all the pure n-dimensional subcom-

plexes of X. The maps f_{ij} are the inclusion maps. $\leqslant j$ iff X_i is contained in X_j. Then $\varprojlim(X,f) = X$.

(2) Let S be an independent set of vertices of P_n. For each point p of S, remove p and the two triangles of $\hat{P}_n$ containing it. Add an edge joining the two end vertices of the polygon, and erect a triangle over that edge. If we do this for each point of S, we get $\hat{P}_n(S)$. We initially construct a map from $\hat{P}_n$ to $\hat{P}_n(S)$ in the case that S is a single point. If 3 is the point of S, then we map as in figure 1. We define the map from $\hat{P}_n$ to $\hat{P}_n(S)$ as the composition of all the maps obtained by removing a point at a time. It is easy to see that the order the points are removed is unimportant.

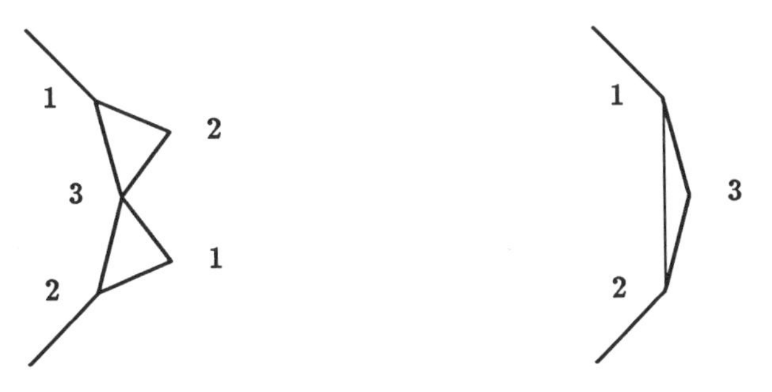

Figure 1

If T is another independent set of $\hat{P}_n$, we say that $S \leqslant T$ iff $S \subset T$. The above gives us a map f_{ST} from $\hat{P}_n(S)$ to $\hat{P}_n(T)$. We claim that $\hat{P}_n = \varinjlim\{\hat{P}_n(S),f_S\}$, where S is non-empty. It is clear that $\hat{P}_n$ maps to every member of this system. Also, no two points of P map to the same point under all the maps f_S. These two observations suffice to show that P is the direct limit.

(3) Suppose that the directed system consists of spaces X_1,X_2,X_3 and maps $f_{13} : X_1 \to X_3$ and $f_{23} : X_2 \to X_3$. $\varprojlim(X,f)$ is a the pullback : the set of all (x_1,x_2) in $X_1 \times X_2$ such that $f_1(x_1) = f_2(x_2)$.

(4) Suppose that the directed system consists of two disjoint complexes X_1 and X_2 and no maps. Its clear that $\lim\limits_{\rightarrow}(X,f) = X_1 \times X_2$ and $\lim\limits_{\leftarrow}(X,f) = X_1 \coprod X_2$. In this case we see that $B(\lim\limits_{\leftarrow}(X,f)) = \lim\limits_{\rightarrow} B(X,f)$. On the other hand,

$B(X_1 \times X_2) = B(X_1) \coprod B(X_2)$ only holds under certain assumptions on X_1 and X_2 (connectivity). Thus we see that it is not always true that $B(\lim\limits_{\rightarrow}(X,f)) = \lim\limits_{\leftarrow} B(X,f)$.

If we take the maps $\lim\limits_{\rightarrow}(X,f) \rightarrow \{X,f\} \rightarrow \lim\limits_{\leftarrow}(X,f)$ and apply B, we get maps $B(\lim\limits_{\leftarrow}(X,f)) \rightarrow (BX,Bf) \rightarrow B(\lim\limits_{\rightarrow}(X,f))$. By the universal property of limits, we get maps

$$B_D : B(\lim_{\leftarrow}(X,f)) \rightarrow \lim_{\rightarrow} B(X,f)$$
$$B_I : \lim_{\leftarrow} B(X,f) \rightarrow B(\lim_{\rightarrow}(X,f))$$

There are simple examples to show that B_I is not always an isomorphism, but we do have

Theorem 1

$B_D : B(\lim\limits_{\leftarrow}(X,f)) \rightarrow \lim\limits_{\rightarrow} B(X,f)$ is an isomorphism.

Proof:

We construct an inverse to B_D. A vertex v of $\lim\limits_{\rightarrow} B(X,f)$ is a compatible collection $\{v_i \mid v_i \in B(X_i)\}$. Write $v_i = g_i^{-1}(p)$ where g_i is a coloring of X_i. Each $g_i^{-1}(p)$ is a collection of vertices of X_i and f_{ij} carries $g_i^{-1}(p)$ into $g_j^{-1}(p)$. Let V be the union of all the v_i's. V is a vertex of B($\lim\limits_{\leftarrow}(X,f)$) and is the desired inverse of v . □

It is also easy to find examples where X = $\lim\limits_{\rightarrow}(X,f)$ and all the X_i are reflexive, but X is not. We do have the following

Lemma 2

If $X = \lim_{\rightarrow}(X,f)$ and for all i, $\Phi : X_i \rightarrow B^2(X_i)$ is one to one, then $\Phi : X \rightarrow B^2(X)$ is one to one.

Proof

Let x and y be any two points of X. we must show that there is a coloring f of X such that $f(x) \neq f(y)$. By the universal property of direct limit, there is an index i such that the x and y project to different points in X_i. Since Φ is one to one on X_i, there is a coloring g of X_i such that $g(x) \neq g(y)$. Composing the projection with g gives the desired coloring f. □

7. Examples

We present several examples of interesting complexes to color.

Triangulations

Suppose that X is a triangulation of the 2-sphere. B(X) is the space of three colorings of X, since X is a 2-complex. Consequently, B(X) is empty unless X is an even triangulation, in which case B(X) is a single triangle. $B(\hat{X})$ is the space of 4-colorings of X, and is well known to be non-empty. We will study the reflexive properties of triangulations in Chapter 5.

The Steiner Complex

Let STS(n) be the simplicial complex whose vertices are all 3-subsets of an n-element set. The top simplices correspond to all the 2-subsets of the set. Given a pair of elements, the top simplex consists of all triples containing that pair. There are $\binom{n}{3}$ vertices, and the complex has dimension n-3. A vertex of B(STS(n)) is a set of triples such that each pair is contained in a unique triple. Such a collection is called a Steiner Triple

System. A well known condition for the existence of a Steiner Triple System is that $n \equiv 1$ *or* 3 (mod 6). It is conjectured that for all such n > 7, B(STS(n)) is non-empty. [Rosa, 1975]

Spreads and Packings

Let X have vertex set the set of all lines of PG(3,q), projective 3-space over the finite field with q elements. For each point in PG(3,q), we make a top simplex out of all the lines passing through the point. X is a (q^2+q)-complex. A vertex of B(X) is a set of lines such that each point lies in exactly one line. Such a set is called a *spread.* A coloring is known as a *packing.*

Ovoids

An ovoid O of PG(3,q) is a set of q^2+1 points, no three of which are colinear. Let the vertex set of X be the planes meeting O. The top simplices correspond to the vertices of O. A vertex of B(X) is a set of planes which cover O and do not meet in O. If a vertex of B(X) contains exactly two planes which meet O in a single point, the vertex is called a *flock.*

Block Designs

A (v,b,r,k,λ) balanced incomplete block design contains v varieties, b blocks, each variety is contained in r blocks, each block contains b varieties, and each pair of varieties lies in λ blocks. Let X have vertex set the blocks of the design. The top simplices correspond to the varieties. If we chose the other order ,since every two varieties lie in some block, there would be no colorings. A vertex of B(X) is called a *parallel class*, a coloring is called a *resolution* or a *parallelism.*

Generalized Quadrangles

Let Q be a generalized quadrangle of order (s,t). Let the vertex set of X be the points of Q, and let the top simplices be the lines of Q. A vertex of B(X) is a set of points such that each line contains exactly one point. Such a set is called an *ovaloid.*

8. Coloring Arbitrary Complexes

It is possible to define B(X) for complexes X which are not necessarily pure. Let X be any complex , and let r be any integer. An r-coloring of X is a map $X \to \Delta^{r-1}$. $B_r(X)$, the space of r-colorings of X , has vertices all $f^{-1}(p)$, where f is an r coloring of X , and p is a vertex of Δ^{r-1} . Every r-coloring of X determines a simplex $\{f^{-1}(p) \mid p \in \Delta^{r-1}\}$ of $B_r(X)$. The new wrinkle is that some of these vertices may be empty , and hence equal . Consequently , $B_r(X)$ is not necessarily pure.

There are three interesting cases. First , suppose X has chromatic number r . That means X has no coloring with fewer than r colors. Define $B_\chi(X) = B_r(X)$. $B_\chi(X)$ is a pure simplicial complex , for if the empty set was a vertex of $B_\chi(X)$, then there is a coloring with fewer than r colors.

Define $B_{\chi+1}(X) = B_{r+1}(X)$. This is also a pure simplicial complex, for if two vertices of a coloring are equal , then they are the empty set , and we have an r-1 coloring .

Finally , define $B_\infty(X) = B_v(X)$, where v = |V(X)| , the number of vertices of X. The vertices of $B_\infty(X)$ are all independent sets of X. Two vertices are adjacent iff they are disjoint. We have the following simple fact

Lemma 1

$$B_\chi(B_\infty(X)) \cong \Delta$$

Proof

Any two vertices of $B_\infty(X)$ which are in the same vertex of $B_\chi(B_\infty(X))$ must intersect , for if they did not , they would be adjacent. If X has vertices $x_1, \ldots x_v$, then these are also a vertex of $B_\infty(X)$. Consequently , every vertex of $B_\chi(B_\infty(X))$ consists of intersecting vertices , and one of them is a single point. It follows that the members of the vertex are all the vertices of $B_\infty(X)$ containing that point. □

Suppose that X is a triangulation of a surface. If X has no three coloring , then $B_4(X) \cong B(\hat{X})$. If X has a three coloring then they are still isomorphic. The vertex consisting of the empty set in $B_4(X)$ corresponds to the vertex $\hat{X} - X$ of $B(\hat{X})$.

The Symmetric Group Complex S_n

As one would expect, the simplicial complex associated to the symmetric group has a rich structure. S_n is reflexive, and we can determine both Hom and B of various spaces involving S_n. Many subcomplexes of S_n have combinatorial interest, such as the complex consisting of all derangements. For even n, we show this is reflexive. The major tools used are various conditions for the extension of a partial Latin square to a Latin square.

1. Basic Properties of S_n

In this section we establish the basic properties of S_n. Recall Δ denotes Δ^{n-1} and S_n is $\mathrm{Aut}(\Delta)$. The vertices of S_n are the permutations of the vertices of Δ and the top simplices are Latin squares. Here are some elementary facts about S_n.

Theorem 1

(a) $\mathrm{B}(\Delta \# \Delta) = S_n$

(b) $\mathrm{B}(S_n) = \Delta \# \Delta$

(c) $B^2(S_n) = S_n$

(d) $\mathrm{Hom}(S_n, \Delta) = 2 \cdot S_n \# \Delta$

(e) $B(S_n \# \Delta) = 2(S_n \# \Delta)$

(f) $\mathrm{Hom}(\mathrm{X}, S_n) = \mathrm{B}(\mathrm{X} \# \Delta \# \Delta)$

Proof:

For (a), $S_n = \mathrm{Aut}(\Delta) = \mathrm{Hom}(\Delta,\Delta) = \mathrm{B}(\Delta \# \Delta)$. (b) follows from [Fisk et al, 1980]. (c) follows from (a) and (b). For (d), $\mathrm{Hom}(S_n,\Delta) = \mathrm{Hom}(\mathrm{B}(\Delta),\mathrm{B}(S_n)) = \mathrm{Hom}(\Delta,\Delta \# \Delta)$ $= 2\mathrm{Hom}(\Delta,\Delta)\#\Delta = 2S_n \# \Delta$. (e) follows from $\mathrm{B}(S_n \# \Delta) = \mathrm{Hom}(S_n,\Delta) = 2S_n \# \Delta$. (f) follows from the calculation $\mathrm{Hom}(\mathrm{X},S_n) = \mathrm{Hom}(\mathrm{X},\mathrm{Hom}(\Delta,\Delta)) = \mathrm{Hom}(\mathrm{X}\#\Delta,\Delta) = \mathrm{B}(\mathrm{X}\#\Delta\#\Delta)$. □

We next use a result of [Sabidussi , 1960] to find the automorphisms of Cartesian products. Define $X^{\#k}$ to be X#X...#X (k factors). Define a graph H to be prime if it can not be written as the cartesian product of two non-trivial factors. Sabidussi showed that the automorphism group of $G_1 \# G_2 \cdots \# G_k$, where each G_i is prime, is equal to the automorphism group of $G_1 \coprod G_2 \coprod \cdots G_k$

If X is a complex, then any automorphism of X is also an automorphism of the underlying graph of X, although the converse is not necessarily true. If X = $X_1 \# \cdots X_k$ where each of the factors is prime, then every automorphism of X can be realized as a combination of permuting copies of factors which are equal and automorphisms of the factors. In particular, if X is prime as a graph, then Aut($X^{\#k}$) has vertices $(f_1, \cdots f_k, \pi)$, where π is a permutation of 1..k and f_i is an automorphism of the i-th copy of X.

We claim that Aut($X^{\#k}$) is k! disjoint copies of $\text{Aut(X)}^{\#k}$. Let π and τ be distinct permutations, and let $(f_1, \cdots f_k, \pi)$ and $(g_1, \cdots g_k, \tau)$ be adjacent vertices. There is an i such that $\pi(i) \neq \tau(i)$. If x = $(x_1, \cdots x_k)$ is a vertex of $X^{\#k}$, then for $f_1(x_1) \# \cdots \# f_k(x_k)$ to be adjacent to $g_1(x_1) \# \cdots \# g_k(x_k)$ all but one coordinate must be equal. If we vary the i-th coordinate of x, under the f's we vary the π(i)-th coordinate, but under the g's we vary the τ(i)-th coordinate. Consequently, we must have $\pi = \tau$. If $(f_1, \ldots f_k, \pi)$ and $(g_1, \cdots g_k, \pi)$ are adjacent then for all but one index i, we have $f_i = g_i$. We thus have the

Theorem 2

If X is prime as a graph, $\text{Aut}(X^{\#k}) = \text{k!} \cdot \text{Aut(X)}^{\#k}$

Corollary 3

(a) $\text{Aut}(\Delta^{\#k}) = \text{k!} \cdot (S_n)^{\#k}$

(b) $\text{Aut}(S_n) = 2 \cdot S_n \# S_n$

Proof:

Δ is prime as a graph, so the Theorem applies. For (b), $\mathrm{Aut}(S_n) = \mathrm{Aut}(\mathrm{B}(S_n)) = \mathrm{Aut}(\Delta \# \Delta) = 2 \cdot S_n \# S_n$. □

In the next section we will interpretate this result in terms of permutations.

Theorem 4

(a) $\mathrm{Hom}(S_n, S_n) = 2 \cdot S_n \# S_n \coprod 2 \cdot B(\Delta^{\#3}) \# \Delta$

(b) If X is 2-connected, $\mathrm{Hom}(X \# S_n) = \mathrm{Hom}(X, S_n) \# 2\Delta \coprod \mathrm{Hom}(X, \Delta) \# 2S_n$

(c) If X is 2-connected, $\mathrm{B}(X \# S_n) = 2\ \mathrm{B}(X \# \Delta) \# \Delta$

(d) $\mathrm{Hom}(S_n \# S_n, \Delta) = 8S_n \# S_n \# \Delta + 4\Delta \# \Delta \# B(\Delta^{\#3})$

(e) $\mathrm{B}(S_n \# S_n) = 4\ S_n \# \Delta \# \Delta$

(f) $\mathrm{B}(S_n \# S_n \# S_n) = 8[\ 3S_n \# S_n + B(\Delta^{\#3})] \# \Delta \# \Delta$

(g) $\mathrm{Hom}(\Delta \# \Delta, S_n) = \mathrm{B}(\Delta^{\#4})$

Proof

The proofs of these results are simple calculations. We will do (a), (b) and (f) . The rest are similar. For (a), $\mathrm{Hom}(S_n, S_n) = \mathrm{Hom}(\Delta \# \Delta, \Delta \# \Delta) = 2\mathrm{Hom}(\Delta, \Delta) \# \mathrm{Hom}(\Delta \# \Delta) + 2\mathrm{Hom}(\Delta \# \Delta, \Delta) \# \Delta = 2S_n \# S_n + 2B(\Delta^{\#3}) \# \Delta$.

For (b), $\mathrm{Hom}(X \# S_n, \Delta) = \mathrm{B}(X \# S_n \# \Delta) = \mathrm{Hom}(X, \mathrm{B}(S_n \# \Delta)) = \mathrm{Hom}(X, 2S_n \# \Delta) = 2\mathrm{Hom}(X, S_n \# \Delta)$ = the right side of (b). For (f), we have $\mathrm{B}(S_n \# S_n \# S_n) = \mathrm{Hom}(S_n, B(S_n \# S_n)) = \mathrm{Hom}(S_n, 4S_n \# \Delta \# \Delta) = 4\mathrm{Hom}(S_n, S_n) \# \Delta \# \Delta + 8\mathrm{Hom}(S_n, \Delta) \# \Delta \# S_n = 4(2S_n \# S_n + 2B(\Delta^{\#3}) \# \Delta) \# \Delta \# \Delta + 8S_n \# \Delta \# 2\Delta \# S_n$ which is the desired result. □

The term $B(\Delta^{\#3})$ is the space of Latin cubes. Next to nothing is known about Latin cubes , so studying this space is quite difficult. Result (g) can be interpreted as saying that Latin squares whose elements are disjoint permutations correspond to 4-

dimensional Latin cubes.

2. Element-wise description of Maps

In the last section we calculated many Hom's and B's. Now, we show exactly what the maps are that make up certain of these spaces.

Theorem 1

(a) All colorings f : $S_n \to \Delta$ are of the form f(σ) $= \gamma\sigma^e(a)$,

where e is 1 or -1, a $\in\Delta$ and $\gamma\in S_n$

(b) All maps f : $S_n \to S_n$ are either

(1) Automorphisms, given by $f(\sigma)=\alpha\sigma^e\beta$, $e = \pm1$, $\alpha,\beta \in S_n$

(2) or essentially colorings, given by $S_n \to \Delta \to S_n$

Proof:

(a) From [Fisk et al 1980], we know that a vertex of B(S_n) consists of all the permutations with a fixed value in a fixed position. Equivalently, this is $\{\sigma|\, \sigma(b) = a\}$ for some a and b in S_n. Since B(S_n) = $\Delta\#\Delta$, there are exactly 2n top simplices of B(S_n) and they are of the form $\{ \{\sigma|\, \sigma(b) = a\} \mid a\in\Delta\}$ or $\{ \{\sigma|\, \sigma^{-1}(b) = a\} \mid a\in\Delta\}$. These two colorings are of the form f(σ) = σ(a) and f(σ) = σ^{-1}(a). Given an arbitrary coloring, it therefore differs from $\sigma^e(a)$ by some automorphism γ of Δ.

(b) Applying B to f gives the map B(f) : $\Delta\#\Delta\to\Delta\#\Delta$. B(f) is either an automorphism or a coloring, and these induce the corresponding maps on S_n. First assume B(f) is an automorphism. We may take B(f) :$\Delta_1\#\Delta_2 \to \Delta_1\#\Delta_2$ to satisfy B(f) = h#g, with h:$\Delta_1\to\Delta_1$ and g:$\Delta_2\to\Delta_2$. We choose $\sigma\in S_n$ and determine the action of f on σ by considering the action of B(B(f)) on the vertex of $\Delta\#\Delta$ corresponding to σ. This vertex is $\{(i,\sigma(i))|\, i\in\Delta\}$. B(B(f)) applied to it is $(h\#g)^{-1}\{(i,\sigma(i))\} = \{h^{-1}(i),g^{-1}\sigma(i))\}$

$=\{(j,g^{-1}\sigma h(j)) | j\in\Delta\}$. Thus we see that f($\sigma$) is $g^{-1}\sigma h$. The case where h and g go to opposite factors is similar. In case that B(f) is a coloring, it is easy to see that (b) holds.

□

We note that Aut(X) is a group preserving its simplicial structure. The map $x \to x^{-1}$ is an automorphism of Aut(X), as are left and right multiplication. Part (b) shows that all automorphisms of S_n are induced by the group structure.

It can be shown by similar arguments that all colorings of $S_n\#S_n$ are one of the two forms

$$S_n\#S_n \to S_n\#S_n \to S_n \to \Delta \tag{1}$$

$$S_n\#S_n \to \Delta\#\Delta \to \Delta \tag{2}$$

where in (1) the first map is an automorphism, the second is group multiplication, and the third is evaluation at some point of Δ. In (2), the first map is evaluation at a point in each coordinate, and the second map is some coloring of $\Delta\#\Delta$.

3. Local connectivity of S_n

In this section we prove that S_n and various subcomplexes of S_n are connected. The main facts needed are properties of disjoint permutations. Two permutations σ and τ are disjoint iff for all x, $\sigma(x) \neq \tau(x)$. Since any two disjoint permutations are contained in a Latin square, [Ryser, 1963], two permutations are disjoint iff they are adjacent in S_n.

Lemma 1

Suppose that we are given r arbitrary permutations of 1...n If $n\geqslant 2r$ then there is a permutation disjoint from all r given permutations.

Proof:

To show the existence of such a permutation, we use Hall's condition for the existence of an SDR [Ryser, 1963]. Let T_i be the set of possible elements for the i-th column of a disjoint permutation. By Hall's Theorem, if there is no SDR, there is a k such that $| T_1 \cdots \cup T_k| \leqslant k-1$, Since each column of the r given permutations has r distinct elements, $| T_i| \geqslant n-r$ and consequently $k-1 \geqslant n-r$. Since the union of the T_i's has at most k-1 elements, these elements must occur in the last n-k columns of each row, so that $n-k \geqslant k-1$. It follows that $2r-1 \geqslant n$ but we assumed that $n \geqslant 2r$. □

In case n=2r-1 there are examples where the result fails. For instance, if n=7 and r = 4 we have

1	2	3	4	5	6	7
2	3	1	5	6	7	4
3	1	2	6	7	4	5
3	1	2	7	4	5	6

In general, take two Latin squares, the first on 1 ... r-1 and the second on r ... 2r-1. Place the smaller one to the left of the larger one, and add any permutation on the left to get r rows. The r columns over the r by r Latin square need r distinct elements, but only 1...r-1 are available.

Define an n-complex to be *t-path connected* if any two n-simplices can be joined by a chain of n-simplices such that adjacent n-simplices intersect in at least t-points.

Theorem 2

If σ is a k-simplex of S_n and n $\geqslant$ 2k + 6t, then Link(σ , S_n) is t-path connected.

Proof:

σ consists of k+1 disjoint permutations of S_n. Let any two (t-1)-simplices α and β be given. By repeated use of the lemma, we can find t mutually disjoint permutations disjoint from the set of k+1+2t permutations determined by α,β, and σ, provided that $n \geqslant 2(k+1+2t)+2(t-1)=2k+6t$. These t new permutations give a simplex γ of

Link(σ,S_n) which is disjoint from both α and β. By the extension theorem for Latin rectangles, there are Latin squares containing $\alpha \bigcup \gamma$ an $\beta \bigcup \gamma$.

Take two simplices A and B in Link(σ,S_n). Choose t-1 points forming α and β in them. The above argument gives us a chain of length 3 : A, $\alpha \bigcup \gamma$, $\beta \bigcup \gamma$, B . □

Define D_n, the derangement complex, to be Link(id,S_n). Equivalently, D_n consists of all the permutations which are adjacent to the identity. This means that $\sigma(x) \neq x$ for all x. Such permutations are called derangements.

Corollary 3

(a) S_n is connected for n > 3

(b) D_n is connected for n $\geqslant$ 2.

(c) S_n is 2-path connected for n > 3.

Proof:

Putting t=1 and k=1, we have D_n is 1-connected if n$\geqslant$ 8. S_n is 2-connected for n $\geqslant$ 8. For the remaining cases, one can check by hand (or better by computer) that they have the specified connectivity. □

S_3 consists of two triangles, and so is not connected. From the description given earlier, it is easy to check that Link(σ , S_4) is not connected for certain 1-simplices σ.

Lemma 4

(a) Suppose that we are given an r by n Latin rectangle. If we choose t columns and place acceptable values in these columns in the r+1-st row, then we can extend this collection to a Latin Square provided n > 2r+t-1.

(b) If n = 2r+t-1 there are counterexamples

Proof:

Without loss of generality, we place i in the i-th column of the r+1-st row for $1 \leqslant i \leqslant t$. Define T_i to be {i} for $1 \leqslant i \leqslant t$ and for i > t let T_i be the set of values not in the i-th column. An SDR for T_i gives an r+1-st row of a Latin rectangle with 1 through t in the desired position. The rectangle can now be extended to a square.

We must show that the union of any k T_i's has at least k members. Assume that we have a minimal collection of T_i's whose union has less than k elements. If none of $T_1, \cdots T_t$ are in the collection then the usual argument works [Ryser, 1963]. Any T_i with i > t contains exactly n-r elements, so we must have n-r < k. Next, the union of any (r+1) T_i's, not including any T_i for $i \leqslant t$, contains exactly n elements. If this were not so, then there is an x not in any of these T_i's, and so it occurs in every column. However, it can occur at most r times - once per row - giving a contradiction. We thus have k-t < r+1. Combined with n-r < k gives us n $\leqslant$ 2r+t-1, contradicting our hypothesis.

Here is a general example. Write down a Latin square on 1 ... (r+t-1) with the last row 1,2...r+t-1 and delete the last t-1 rows. To the right add any r by r Latin square on r+t ... 2r+t-1. If we place 1 through t in the first t columns of the r+1-st row, we can not extend to a Latin Square. In the last r columns of the r+1-st row, we can use only the r-1 elements t+1 ... r+t-1. □

For example with r=3, t=2, n=7, we have

2	3	4	1	5	6	7
3	4	1	2	6	7	5
4	1	2	3	7	5	6
1	2					

It is surprising that the prescription of just two elements leads to the doubling of the size of the Latin square.

Corollary 5

X = $\{\sigma \in D_n | \sigma^2(1) \neq 1\}$ is a connected subgraph of D_n for n > 7.

Proof:

Let σ , $\tau \in X$. Consider the three permutations σ , τ , identity and label two columns of the fourth row so that any permutation γ extending the labeling would satisfy $\gamma^2(1) \neq 1$. For instance, set $\gamma(1) = a$, $\gamma(a) \neq 1$ for some appropriate a. By Lemma 4, there is a permutation γ disjoint from σ,τ,identity with $\gamma^2(1) \neq 1$. Consequently, $\sigma\gamma$ and $\gamma\tau$ are edges of S_n and so X is connected. □

The next result is used in the proof of the reflexivity of D_n.

Corollary 6

Y = $\{\sigma \in D_n | \sigma^2 \neq \text{identity}\}$ is a connected subgraph of D_n for n > 7.

Proof:

Let $X_p = \{\sigma \in D_n | \sigma^2(p) \neq p\}$. The above corollary shows that X_p is connected for all p. Since the union of the X_p is Y, it suffices to show that X_p is connected to X_q. Suppose that $\sigma^2(p) \neq p$ and $\tau^2(q) \neq q$. Consider the three permutations σ,τ,identity. In the fourth row put an a over position 1 and 1 anywhere but in position a, such that 1 and a are possible column values. Since $n>7$ this is possible. There is a γ disjoint from all three of these permutations. By construction $\gamma^2(1) \neq 1$ and so $\gamma \in X_1$. We thus have that X_1 is joined to both X_p and X_q . □

If n is odd, there is no derangement σ satisfying $\sigma^2 = 1$, so Y $= D_n$ is connected by Corollary 3.

4. The Derangement Complex

D_n is the link of the identity in S_n. Since S_n acts transitively on its vertices, D_n is isomorphic to the link of every vertex of S_n. In this section we calculate B(D_{2m}), show that D_{2m} is reflexive, and determine its automorphisms.

Suppose that σ is an r-simplex of S_n. σ is a set of permutations $\sigma_0 \cdots \sigma_r$ which are all disjoint. We identify σ_i with the set of points $\{(j,\sigma_i(j)) | \ j \in \Delta\}$ in $\Delta \# \Delta$. We denote by $\Delta \# \Delta - \sigma$ the subcomplex of $\Delta \# \Delta$ whose vertices are none of $(j,\sigma_i(j))$ for any i and j. Since σ is an r-simplex, each row and column of $\Delta \# \Delta$ has r+1 elements in σ, so $\Delta \# \Delta - \sigma$ is an (n-r-2)-complex, with 2n top simplices, and $n^2 - n(r+1)$ vertices.

Suppose that f:$\Delta \# \Delta - \sigma \to \Delta^{n-r-2}$ is a coloring. If we color alike all the vertices corresponding to the same permutation in σ, then f extends to a coloring $\tilde{f}$ of $\Delta \# \Delta$. Considered as a simplex of S_n, $\tilde{f}$ can be written $\sigma * \tau$ for some simplex τ. We thus have a map B($\Delta \# \Delta - \sigma$) $\to$ Link(σ,S_n). Conversely, every coloring of the right hand side corresponds to one of the left, so we have that B($\Delta \# \Delta - \sigma$) $\cong$ Link(σ,S_n).

We conjecture that B(Link(σ,S_n)) $\cong \Delta \# \Delta - \sigma$ for n sufficiently large relative to r. For r=0, we have shown it to be true. We will show it to be true for r=1 and n even. There are a few computer calculations which support this conjecture for r=2 and r=3. For example, B(Link(σ,S_n)) $\cong \Delta \# \Delta - \sigma$ in S_6 for σ equal to {id,(123)(456)} and {id,(1243456),(135)(246)}.

Theorem 1

D_n is reflexive for n even and greater than 6, or n=4.

Proof:

The idea is to find a large reflexive subcomplex of D_n and then show that colorings extend uniquely from it to D_n. Suppose n=2m. The complex $L(K_n)$ has as vertices all the edges of the complete graph K_n and has top simplices corresponding to the n vertices of K_n. See Chapter 5 for more details about $L(K_n)$. All we need here is that $L(K_n)$ is reflexive [Fisk et al, 1980]. A vertex of B($L(K_n)$) is a set of m edges of K_n which have no vertex of K_n in common.

Let id denote the identity permutation. Define the map $\tilde{\alpha} : \Delta\#\Delta - id \to L(K_n)$ by sending (i,j) to the edge joining i and j. This is clearly a non-degenerate map. Applying B gives us a map $\alpha{:}B(L(K_n)) \to D_n$. The vertex $\{(a_1,b_1) \cdots (a_m,b_m)\}$ of $B(L(K_n))$ is sent to the permutation $(a_1 b_1) \cdots (a_m b_m)$ of S_n. Note that this is only well defined for n even. α is one to one, so we consider $B(L(K_n))$ imbedded in D_n as $\mathrm{Im}(\alpha)$.

A coloring G of $B(L(K_n))$ is determined by some vertex p of $L(K_n)$. $G((a_1,b_1) \cdots (a_m,b_m))$ is the other element in the transposition containing p. If F is a coloring of D_n, then without loss of generality $F|\mathrm{Im}(\alpha)$ is given by $F((a_1 b_1) \cdots (a_m b_m))$ equals the other element in the transposition containing 1. Equivalently, $F(\sigma) = \sigma(1)$ for $\sigma \in \mathrm{Im}(\alpha)$.

We next show that for any σ in D_n, $F(\sigma)$ is either $\sigma(1)$ *or* $\sigma^{-1}(1)$. This holds for σ in $\mathrm{Im}(\alpha)$. Let σ have a cycle representation (1a...b)(...)...(...). It is easy to find an X in $\mathrm{Im}(\alpha)$ such that σX has no fixed points. In fact, we can find an X such that X contains (1c), for any c distinct from a and b. It follows that X is adjacent to vertices colored every color except a and b. Consequently, $F(\sigma)$ is $\sigma(1)$ or $\sigma^{-1}(1)$.

If σ and τ are in $D_n - \mathrm{Im}(\alpha)$, then either $F(\sigma) = \sigma(1)$ and $F(\tau) = \tau(1)$ or $F(\sigma) = \sigma^{-1}(1)$ and $F(\tau) = \tau^{-1}(1)$. Assume that $F(\sigma) = \sigma(1) = a$ and $F(\tau) = \tau^{-1}(1) = b$. By Lemma 3.4, for $n > 7$ there is a permutation γ disjoint from τ, σ and the identity with b in the first column and a in the b-th column. γ is thus adjacent to σ and τ and is in D_n. However, $F(\gamma)$ is either a or b which is impossible.

By Corollary 6 of the last section, $D_n - \mathrm{Im}(\alpha)$ is connected, and so every coloring is of the form $F(\sigma) = \sigma(a)$ or $\sigma^{-1}(a)$ for some a. These are also the colorings of S_n. The difference is that $\sigma^{\pm 1}(a)$ never takes on the value a, so the complex is $\Delta\#\Delta - id$. Consequently, $B^2(D_n) = B(\Delta\#\Delta - id) = D_n$. □

Now that we know that D_{2m} is reflexive, we can determine its automorphism group.

Corollary 2

(a) $\mathrm{Aut}(D_{2m}) \cong 2{\bullet}S_{2m}$

(b) All automorphisms of D_{2m} are given by $\sigma \to \alpha\sigma^e\alpha^{-1}$, for fixed $\alpha \in S_{2m}$ and e = ± 1.

Proof:

Since D_{2m} is reflexive, we know $\mathrm{Aut}(D_{2m}) = \mathrm{Aut}(\mathrm{B}(D_{2m})) = \mathrm{Aut}(\Delta \# \Delta - id)$. Suppose that f is an automorphism of $\Delta \# \Delta - id$. Clearly either all rows go to rows or all columns go to columns. By Lemma 3, f extends to an automorphism F of $\Delta \# \Delta$. From Corollary 2.1.3a, F(x,y) is either $(\sigma(x),\tau(y))$ or $(\tau(y),\sigma(x))$ for some σ and τ. Since F carries the diagonal of $\Delta \# \Delta$ to itself, we see that $\sigma = \tau$, so f(x,y) is either $(\sigma(x),\sigma(y)$ or $(\sigma(y),\sigma(x))$. For any σ and τ we can find an x and y such that $\sigma(x) \neq \tau(y)$ and $\sigma(y) \neq \tau(x)$, so the automorphism complex is isomorphic to $2S_n$. The induced automorphism on D_{2m} is as given in (b). □

Lemma 3

If f is an automorphism of $\Delta \# \Delta - \sigma^{r-1}$, and n $\geqslant$ 2r+1, then f extends uniquely to an automorphism of $\Delta \# \Delta$.

Proof:

Let $r_1, \cdots r_n$ be the rows of $\Delta \# \Delta - \sigma$, and assume that f carries r_1 to r_i. There are r positions in the i-th row which are in σ. It is possible for r rows to be sent to columns, but no more. If a row is sent to a column, then by the same argument, there can be at most r rows sent to rows, and so 2r $\geqslant$ n. We conclude that r sends rows to rows and columns to columns.

Define f(a,b) to be the intersection of the row containing f(a-th row) and the column containing f(b-th column). This f agrees with the the original automorphism in $\Delta \# \Delta - id$, and so is the desired extension. This extension is clearly unique. □

In the case n=2r, there is a counterexample. Partition $\Delta\#\Delta$ into four r by r Latin squares A,B,C,D, where A and D are Latin squares on 1...r and B and C are Latin squares on r+1...2r. Let the simplex γ have vertices $f^{-1}(c)$, for c = r+1 ... 2r. $\Delta\#\Delta-\gamma$ has as vertex set $A\bigcup D$, and so $\mathrm{Aut}(\Delta\#\Delta-\gamma) = \mathrm{Aut}(A) \times \mathrm{Aut}(D)$. In particular, there are automorphisms which send rows of A to rows of A and columns of D to columns of D.

As opposed to $B(S_n\#\Delta) \cong 2S_n\#\Delta$, we have

Corollary 4

If n is even, $B(D_n\#\Delta^{n-2}) \cong 2n{\bullet}S_n$

Proof:

$B(D_n\#\Delta^{n-2}) \cong \mathrm{Hom}(\Delta,B(D_n)) \cong \mathrm{Hom}(\Delta,\Delta\#\Delta-id)$. Suppose that $f_1 \cdots f_{n-1}$ is a simplex of $B(D_n\#\Delta)$. For each x, $\{f_i(x)|\ 1\leqslant i\leqslant n-1\}$ is a simplex of $\Delta\#\Delta-id$. It's clear that all f_i map to rows (or all to columns), for if f_1 went to a row and f_2 to a column, some some $f_1(x)$ would not be in the same row or column as $f_2(x)$. If f_1 and f_2 are in different rows, then all the f_i are in different rows. However, $\{f_1(x)|\ x\in\Delta^{n-2}\}$ has n-1 elements, and there are n rows. Consequently, all f_i map to some fixed row or column. The set of all maps to a single line is S_n , and there are 2n lines. □

If we knew that $B(\mathrm{Link}(\sigma^{r-1},S_n) \cong \Delta\#\Delta-\sigma$, then $B(\mathrm{Link}(\sigma,S_n)\#\Delta^{n-r-1})$ $\cong \mathrm{Hom}(\Delta,\Delta\#\Delta-\sigma)$. As above, all these maps are to fixed lines, so we would get as a result $2n{\bullet}S_{n-r}$.

We can talk about derangements in any complex composed of endomorphisms. Recall that the n-dimensional octahedron is realized as $(S^0)^{*n}$. The endomorphims form a complex isomorphic to $S_n[2^{2n}{\bullet}\Delta^0]$, since each endomorphism determines a permutation, and for each permutation there are 2^{2n} different choices. The derangements are the endomorphisms adjacent to the identity, and is isomorphic to $D_n[2^{2n}{\bullet}\Delta^0]$.

5. General Decomposition and the Oberwolhfach Problem

It is possible to expand the idea of the coloring functor to apply to more general decompositions. Very generally, suppose we can decompose an object O into pieces $\mathbf{p}_1 \cdots \mathbf{p}_k$ in many different ways. We can describe the global structure of all these decompositions as follows. Let B(O) have as vertices all the pieces $\mathbf{p}_i$ which occur in some decomposition. A set of pieces forms a simplex of B(O) iff they are all in the same decomposition.

Examples

(1) The objects are pure simplicial complexes . The decomposition is determined by a coloring, and the pieces are the vertices of B(-).

(2) The objects are regular graphs . The decomposition is a coloring of the edges, and the pieces are 1-factors.

(3) The objects are arbitrary graphs G. The pieces are subgraphs of G isomorphic to a fixed graph H. A collection of subgraphs isomorphic to H is a decomposition if they are edge disjoint, and every edge of G is in exactly one member of the collection. (2) is the special case where H is a set of disjoint edges.

(4) The objects are directed graphs. A piece is a subgraph isomorphic to some directed graph H. Decomposition is edge disjoint union.

(5) The object are subsets of squares of an infinite checkerboard. The pieces are isomorphic copies of a fixed polyomino. Decomposition is square-disjoint union.

We are going to investigate the following special case. We wish to decompose the directed complete graph K_n* into copies of the graph H which is the disjoint union of directed cycles $C_1 \cdots C_s$. We assume cycle C_i has length n_i and $n_1+\cdots+n_s = n$. The directed Oberwohlfoch problem asks whether this can be done.

We can find at most n-1 disjoint copies of H in K_n*, for there are n-1 edges going into any vertex of K_n*. If we view each copy of H in a decomposition as the permutation which is the product of the cycles of H viewed as permutations, then the directed Oberwohlfoch problem asks if there is an n-2 simplex in S_n whose n-1 vertices all have the cycle type $(n_1, \cdots n_s)$. This leads us to the definition : $\tilde{S}(n_1, \cdots, n_s)$ is the subcomplex of S_n whose vertices all have cycle type $(n_1, \cdots, n_s)$. We can ask for the dimension of this complex. If the dimension is r, define $S(n_1, \cdots, n_s)$ to be the subcomplex of S_n consisting of all r-simplices whose vertices all have cycle type $(n_1, \cdots, n_s)$. We then ask if $S(n_1, \cdots, n_s)$ is reflexive.

Examples

(1) The vertices of S(2,2,...2) are all decompositions of S_n into n/2 disjoint edges. We have seen that this is $B(L(K_{n,n}))$, which has dimension n-2 and is reflexive.

(2) S(n) has vertices all directed cycles of length n in K_n*. It is known [Sotteau, 1980] that the dimension of S(n) is n-2. A few of the small S(n)'s have been shown to be reflexive. For instance, S(5) has 24 vertices and is isomorphic to $6 \bullet \Delta^3$. In $\tilde{S}(5)$, there are triangles which are not contained in any tetrahedron of S(5). This shows that the $\tilde{S}(n_1, \cdots, n_s)$ are not necessarily pure simplicial complexes.

(3) Every nk-2 simplex of S(k,k,...k) (n k's) corresponds to a decomposition of $K_{nr}*$ into directed k-cycles. Such a decomposition would follow from a decomposition of K_n into k-cycles. It is conjectured [Sotteau, 1980] that this happens for n and k odd, k > 1.

(4) S(2,3) is the subcomplex of S_5 of all permutations of the form (ab)(cde). Figure 1 below shows that S(2,3) $\cong \Delta^4 \# \Delta^4 - id$. The permutations in any row or any column are adjacent, and there are no other adjacencies. If we let f(i,j) be the permutation in the i-th row and the j-th column, the f(i,j) contains the transposition (i,j). Furthermore, $f(i, j)^{-1} = f(j,i)$.

*	(21)(345)	(31)(254)	(41)(235)	(51)(243)
(12)(354)	*	(32)(145)	(42)(153)	(52)(134)
(13)(245)	(23)(154)	*	(43)(125)	(53)(142)
(14)(253)	(24)(135)	(34)(152)	*	(54)(123)
(15)(234)	(25)(143)	(35)(124)	(45)(132)	*

Figure 1

Complexes Arising From Geometry

In this chapter we study the colorings of various simplicial complexes which arise from geometry. We will find that under certain conditions, B(X) has a geometric interpretation. We begin with the richest collection of complexes, those determined by points and lines in the plane. Every configuration of points in the plane has a dual configuration, and the colorings of the complex associated to the original configuration is sometimes isomorphic to the dual configuration. After a brief look at spreads, we investigate the complex associated to the hyperbolic surface in projective 3-space. Finally, studying hermitian surfaces in 2 and 3 dimensional projective space, we find an infinite family of complexes which might be self dual. A basic reference for the projective geometry used in this chapter is [Hirshfeld,1979].

1. Points and Lines in the plane

In this section we investigate various complexes which arise from points and lines in the projective plane. Suppose that L is a collection of lines in a finite projective plane. The plane needn't be Desarguesian, but all of our examples will be from PG(2,q) for some q. We associate a simplicial complex to L by taking the vertices to be the points lying on a line of L. For each line of L we get a top simplex whose vertices are all the points contained in L. We will call this simplicial complex L as well. If each line has q+1 points on it, then a coloring of L uses exactly q+1 colors. A vertex of a coloring of L is a set S of points in the plane such that there is exactly one member of S on every line of L. We can construct colorings of L quite easily. Let p be a vertex of the plane which is not on any line of L. For any line which passes through p, color alike all the vertices of L which meet the line. No two points on a line of L are colored alike, so this gives us a valid coloring.

In order to use this construction, we construct another complex. Define the complementary complex L^c to L as follows. The vertices are all the lines which pass through a vertex not in L. Each vertex which does not lie in L determines a top simplex : all those lines passing through it. The preceding paragraph showed the existence of the map

$$\beta : X^c \to B(X)$$

The remainder of this section will be devoted to a study of β. In particular, we will find that it is sometimes an isomorphism.

By duality, there is also a map from X to $B(X^c)$, which we also call β. We send each point of X to the set of all lines passing through it. This is the same mapping as we get from the composition $X \to B^2(X) \to B(X^c)$. Our first result determines when β is injective.

Lemma 1

Let L be a set of n lines of PG(2,q).

(1) If the lines of L do not all meet in one vertex, then $\beta : L^c \to B(L)$ is 1-1.

(2) If the points of L^c are not colinear, then $\beta : L \to B(L^c)$ is 1-1.

Proof:

A vertex of L^c is a line p meeting some point not on L. $\beta(p)$ consists of all points of L on p. Suppose the line q satisfies $\beta(p) = \beta(q)$. If there are two points of L on p, then this line also contains q and so p equals q. If not, then L meets p in just one point and so the lines of L are concurrent. (2) follows from (1) by duality. □

If a set has very many lines, then there are so many restrictions on the existence of a coloring that we should expect that the only colorings are those determined by the complement. Our next result is an example of this.

Theorem 2

Let S be a set of n points of PG(2,q), and let X denote the associated simplicial complex. Then

(1) If $n \leqslant \sqrt{q}$ then $\beta : X \to B(X^c)$ is bijective on vertices.

(2) If $n \leqslant \sqrt{q}$ and all points of S are contained in a line, then β is a isomorphism.

(3) If $n = 1+\sqrt{q}$ and all points of S are contained in a line, and are not contained in a line of some Baer subplane, then $\beta : X \to B(X^c)$ is an isomorphism.

Proof:

We begin with (1). Choose a vertex F of $B(X^c)$. It suffices to show that F is contained in a line. If n=1, then X^c consists of all lines in the plane not passing through a specified point. Since any two points not colinear with this specified point are adjacent in $B(X^c)$, there is exactly one coloring, and so β is an isomorphism.

If not all points of S lie on a line, we show that $|F| \geqslant q+2-n$. Recall F is a set of points, exactly one of which is on each line missing S. Let p be a point not in F nor in S, lying on a line joining two points of S. If no such point exists, then any line joining two points of S contains only points of S and F on it. Since any line missing these points of S will intersect these points of F, it follows that all the points of S lie on the line. Since p lies on a line joining two points of S, there are at most n-1 distinct lines through p meeting S. Consequently, there are at least q+1-(n-1) lines through p not meeting S, and every one of these contains a point of F.

Suppose that all the points of S lie on a line T. If p is a point of T not in S, then any of the n lines through p distinct from T contain at most one point of F, so $q \geqslant |F|$. Since there is at least one point of F on each of these lines through p, $|F| = q$.

Now assume that not all points of F lie on a line. We will show that $|F| \leqslant n(n-2)+1$. Let p and r be points of F. If a line contains two points of F, then it must contain a point of S. Let the line pr meet S in the point w. Let u be a point of F not on the line pr. Considering all the lines from u to points of F on pr, we see there are at most n-1 points of F on the line pr. At most n lines through p contain points of F, so there are at most 1+n(n-2) points in F. Consequently, $1+n(n-2) \geqslant |F| \geqslant q+2-n$ or $n^2 > q+1$. If the points of S are colinear, then 1+n(n-2)$\geqslant$q, so n$\geqslant\sqrt{q}+1$. In either case, if $n \leqslant \sqrt{q}$ then we can not have such a vertex. From Lemma 1, β is 1-1, so we have shown that β is bijective.

For part (2), we know all the vertices of $B(X^c)$. Suppose we have a collection of vertices of $B(X^c)$ which form a simplex, but the simplex is not in the image of β. As S is contained in a line, a vertex is all the points of a line through a point of S, other than that point of S. If two of these vertices are distinct, they must pass through the same point of S. Consequently, β is a isomorphism.

Suppose that $n = 1+\sqrt{q}$. We claim that $F \bigcup S$ is a Baer subplane. We must show that between any two points of $F \bigcup S$ there are exactly $\sqrt{q}+1$ lines which meet $F \bigcup S$, and they each contain $\sqrt{q}+1$ points of $S \bigcup F$. Let F be a vertex of $B(X^c)$ and assume that S is contained in a line L. We saw above that |F|=q. This implies that every line through two points of F contains one point of S and exactly $\sqrt{q}$ of F. Since there are n distinct lines through a point of F which meet S, it follows that the line between a point of F and one of S contains the correct number of points. Finally, consider two points of S. They lie on the line L, which contains $\sqrt{q}+1$ points. Hence, $S \bigcup F$ is a Baer subplane.

The same argument for part (2) shows that β is an isomorphism, not just a bijection on vertices. □

In the case that $n = \sqrt{q}+1$ and S is contained in the line of a Baer subplane, we do not have an isomorphism . Any Baer subplane intersects any line, so it follows that the subcomplex Im(β) in $B(X^c)$ meets the subcomplex of colorings whose vertices are all Baer subplanes in the vertex corresponding to the unique line through S. We will construct colorings involving Baer subplanes in the next section.

Recall that a oval of a projective plane of odd order q consists of q+1 points, no three colinear. If q is even, there are q+2 points, no three on a line.

Theorem 3

Suppose that K is a subset of the points of an oval C contained in PG(2,q). Furthermore, assume that q is even and greater than 2 , and $|K| \neq q+1$. If X denotes the simplicial complex whose top simplices are the points of K and vertices the lines incident with them, then

$$\beta : X \to B(X^c) \quad \text{is an isomorphism}$$

Proof:

X^c has as vertices all points lying on lines which miss K. A vertex of $B(X^c)$ is a set F of points such that every line not meeting K contains exactly one point of F. If two points of C-K are contained in F, then the line joining them is a secant of C, and therefore is external to K. Since no external line contains two points of F, we see that F contains at most one point of C-K.

Assume first that F contains no point of C. We claim that F has q-1 members. There are $q(q-1)/2$ external lines to C. Through an external point of C, there are (q+2)/2 secants of C, so there are q/2 external lines through p. This gives $|F|(q/2) = q(q-1)/2$, so $|F| = q-1$. Every line through two points of F must meet K and since q is even, such a line is a secant of C. Thas [Thas,1977] has shown that such a set of q-1 points must be all the external points of a secant of C. Thus F lies on a line, and so is in the image of β.

Next, we assume that F contains exactly one point p of C-K. The set F' = F-p has the property that every line external to C meets F' in a unique point. By the above, F' consists of all the external points of a secant L of C. Suppose that L does not pass through p. Since $|K| \neq q+1$, there is another point r of C-K. L does not contain r, for there are q/2 lines through r missing C, and F would not intersect any of them. Thus, r is not on L, and pr is external to K. This is a contradiction, for then the line pr contains two points of F, namely p and the intersection of pr and L. Consequently, in all cases L is in the image of β.

To see that β is injective, let L and M be lines passing through points of K. Clearly if L and M meet every external point in the same point, then L equals M.

To show β an isomorphism, consider a collection of vertices forming a coloring of X^c. This is a collection of q+1 lines passing through points of K. If three of the lines are not coincident, then their intersections with another of the lines gives the impossible configuration of three colinear points in a conic. If there are only three lines, then q is 2.

□

Remarks

(1) In case q is 2 and K is C minus a point, the complement X^c has exactly one line in it. $B(X^c)$ has one top simplex, while X has four.

(2) In general, if K is C minus a point p, β is not an isomorphism. X^c consists of all the lines missing C-p, and their incident points. But, if a line meets one point of C (recall q is even and there are no tangents), it meets another point of C. Thus the complement of C is the same as the complement of C-p. Therefore $B(X^c)$ is the simplicial complex associated to C and not to C-p.

(3) In the case q is odd, we still believe the theorem to be true. It has been computationally verified up to q=7. Attempting to copy the proof, we still get that $|F|$ is q or q-1. The case $|F| = q-1$ follows by the same argument as in [Thas and Fisher 1979]. We need a result that says that if a set of q external points of a conic meet every external line, then they consist of all the external points on a tangent or secant.

2. Baer Subplanes

In this section we discuss some questions concerning colorings of certain simplicial complexes involving Baer subplanes. We will investigate two complexes : one of them has colorings, but we do not know their structure, and the other has correct parameters for the existence of colorings, but only potential vertices of a coloring are known to exist.

First, consider the simplicial complex whose vertices are all the Baer subplanes of PG(2,q), q a square. The top simplices correspond to the points of PG(2,q), and consist of all the Baer subplanes containing a fixed point. A vertex of a coloring of this complex is a collection of Baer subplanes such that every point of PG(2,q) occurs exactly once in each subplane. Such decompositions exist [Bruch,1960], and are easily constructed using a cyclic projectivity. The number of Baer subplanes required is $(q^2+q+1)/(q+\sqrt{q}+1) = q-\sqrt{q}+1$. The total number of Baer subplanes is $q(q-\sqrt{q}+1)(q+1)(\sqrt{q}+1)$. If there were a coloring, the number of colors would have to be $q(q+1)(\sqrt{q}+1)$.

More interesting is the complex arising in Theorem 2 of the previous section. Let S be the set of all points of a line of some Baer subplane of PG(2,q). Let X be the simplicial complex whose top simplices are the points of S, and let X^c be its complement. We construct colorings of X^c whose vertices include all Baer subplanes, and the line through S, and hence $B(X^c) \neq X$.

If B is a Baer subplane containing S, then B-S is a set of points such that the line through any two of them meets S. Thus any Baer subplane containing S gives a potential vertex of $B(X^c)$. Suppose that we can find q Baer subplanes such that any two of them intersect exactly in S. The q points not in S of one such Baer subplane are disjoint from the q points not in S of any other such subplane. Since there are q^2 points not on the line containing S , every point of PG(2,q), other than those on the line containing S, is contained in exactly one of the Baer subplanes. This gives us a coloring of $B(X^c)$: one vertex is the points on the line containing S, other than S. The other vertices are the q Baer subplanes.

Here is a explicit construction of one such decomposition. Let B(r,s) be the subplane of PG(2,q) generated by all linear combinations over $GF(\sqrt{q})$ of (0,1,0) , (0,0,1) , $(1,s\alpha,r\alpha)$, where $\alpha \in GF(q)-GF(\sqrt{q})$. Let the points S be all points on the line generated by (0,1,0) and (0,0,1). A point of B(r,s) not on this line has the form $(1,s\alpha+a,r\alpha+b)$, where $s,r,a,b \in GF(\sqrt{q})$. If B(r,s) and B(u,v) intersect outside of S, then it is easy to see that r = u and s = v. The set of q B(r,s), for r,s in $GF(\sqrt{q})$ form the desired set of Baer subplanes.

The simplest case is when q is 4. A computer calculation showed that $B(X^c)$ had 25 top simplices and 24 vertices, whereas X has 3 top simplices and 13 vertices. Furthermore, X^c has 18 vertices and 8 top simplices, while $B^2(X^c)$ has 17 vertices and 8 top simplices , and $B^3(X^c) \cong B(X^c)$.

We can't prove that $B^3(X^c) \cong B(X^c)$, but we can show why there were only 17 vertices in $B^2(X^c)$ for q=4. Let $\Phi:X^c \to B^2(X^c)$ be the natural map. Let the set S be contained in a line L. Under Φ, all the points of L-S are sent to the same point. Recall that $\Phi(x)$ is all the vertices of $B(X^c)$ which contain x. Now every vertex of $B(X^c)$ is either a line through a point of S or a Baer subplane containing S. The only line passing through a point of S and containing a point of L-S is the line L. Thus Φ takes all the points of L-S to one point, the vertex consisting of L alone. In case q is 4, there were two points in L-S, and they collapsed to one.

3. Spreads in PG(3,q)

In this section we briefly discuss spreads in PG(3,q). Let X be the simplicial complex whose vertices are all the lines of PG(3,q), and whose top simplices correspond to the points of PG(3,q). A vertex of B(X) is called a spread ; a coloring of X is called a packing. It is well known that X has colorings [Denniston ,1972].

Let us look at an example. If q is 2 then X has 15 top simplices and 35 vertices. A computer calculation shows that B(X) has 56 vertices and 240 top simplices. B(B(X)) is not isomorphic to X but instead has 30 top simplices and 35 vertices.

In general, B(B(X)) is not isomorphic to X. We obtain a vertex of B(X) by choosing a line of PG(3,q) and taking all the spreads which contain it. A coloring of B(X) is obtained by taking all the vertices corresponding to lines passing through a fixed point. There is however, another way to get colorings, and that is to take all the vertices corresponding to lines contained in a fixed plane.

Here is another way of seeing this. Let π be a polarity of PG(3,q) taking points to planes, planes to points, and lines to lines. π takes spreads to spreads. Applying π to the coloring determined by a point p gives a coloring determined by the plane $\pi(p)$.

Define Y to be the complex whose vertices are the lines of PG(3,q) and whose top simplices are the points and the planes of PG(3,q). In case that q is 2, we see that B(B(X)) is Y. In general Y $\to$ B(B(Y)) is 1-1, and we wonder if it is an isomorphism.

4. The Hyperbolic Quadric in PG(3,q)

In this section we study the colorings of a simplicial complex determined by the lines lying in a certain algebraic variety. The locus H of the equation

$$x^2 + y^2 + z^2 + w^2 = 0$$

in PG(3,q) is called the hyperbolic quadric. It contains $(q+1)^2$ points and $2(q+1)$ lines. The lines split into two sets of size q+1, A and B, called rulings, such that no two lines of A (resp. B) intersect, but every line of A intersects every line of B. Let X be the simplicial complex whose vertices are the points of H and whose top simplices correspond to the lines lying on H.

X has colorings, and one construction of them is as follows. Let L be a line of PG(3,q) disjoint from H, and let P be a plane through L. P intersects every line of H in exactly one point, and so the intersection of P with H is a vertex of B(X). The q+1 planes through L determine a coloring of X.

We construct a complement to X. Let X^c have as top simplices all the lines missing H, and as vertices all the planes through these lines. The above construction gives a map $\beta : X^c \to B(X)$. By duality, there is also a map $\beta : X \to B(X^c)$. In this case a point of X is sent to all the planes of X^c which contain it.

Before we proceed, we must see that there actually are lines disjoint from H. A plane is a tangent plane of H iff it contains a line of H. A non-tangent plane contains no line of H. The intersection of a non-tangent plane and H is a conic, and since it contains no lines, the conic is non-degenerate. A plane contains q(q-1)/2 lines which do not intersect the conic and therefore miss H as well. The number of non-tangent planes is $(q^3+q^2+q^1+1)-(q+1)^2 = q^3-q$. Let there be N external lines. Counting the pairs (L,P), where L is an external line to H and P is a plane containing it, we see $N(q+1) = (q(q-1)/2)(q^3-q)$ so N is $q^2(q-1)^2/2$. Consequently, X^c has q^3-q vertices, $q^2(q-1)^2/2$ top simplices, and every vertex is in (q-1)q/2 top simplices. In particular, X^c is not empty.

For q greater than 2, it is not true that $B(X) \cong X^c$. But we do have

Theorem 1

If q is even, $\beta : X \to B(X^c)$ is an isomorphism.

Proof:

It is convenient to change our point of view. Let π be the polarity defined by H. Points of H go to tangent planes through that point, and lines of H are fixed by π. Applying π to X^c, we still get external lines, but the planes through these lines go to the points on these lines. We consider X^c consisting of the external lines to H and the points on these lines. β takes a point h of H into the set of all planes of X^c containing h. Applying π, we consider β to take h to the set of all points of the tangent plane at h not in H.

Now let S be a vertex of $B(X^c)$. This is a set of points of PG(3,q)-H such that every external line of H contains exactly one point of S. Let L be an external line and let P be a plane through L. $P \cap H$ is a conic and $P \cap S$ is a set of points, exactly of of which is on every line external to the conic. Now by the same argument as in Theorem 3.1.3 ,$P \cap S$ consists of all the points not in $P \cap H$ which lie on a tangent or secant line of $P \cap H$. Call the line M and let $L \cap M = p$. Since any plane containing L contains p, each of the q+1 planes through L gives a line passing through p. In particular, the line between p and any other point of S contains at most two points of H. Thus any two points of S are joined by a line N with all but one or two points of N in S. Thus from 3 non-colinear points of S we see that the plane less points of H determined by them is in S. This accounts for all the lines, so S is a plane minus H.

S must be a tangent plane, for otherwise some external line would contain more than one point of S. Finally, we check the top simplices. Consider a collection of points $p_1, \cdots, p_{q+1}$ of H such that $\pi p_1, \cdots, \pi p_{q+1}$ disjointly cover PG(3,q)-H. If p_1 and p_2 are not a line of H, then the line $\pi(p_1 p_2)$ is covered by both πp_1 and πp_2. Thus, any two points lie on a line of H, and so are all colinear. □

Remarks

As a complex, X is isomorphic to $\Delta^q \# \Delta^q$. B(X) is therefore isomorphic to S_{q+1}. Clearly this is not X^c, for S_{q+1} has (q+1)! vertices and X^c has only $q^3 - q$. In case q=3, X^c is isomorphic to S_4. We will see in Chapter 5 that with the right kind of restrictions, β is sometimes an isomorphism.

5. Hermitian Varieties

In this section we investigate properties of the Hermitian varieties in $PG(2,q^2)$ and $PG(3,q^2)$. We show that a certain complex associated with the Hermetian variety in PG(2,q^2) might be self dual. We also establish the existence of a coloring for the complex associated to the Hermetian variety in PG($3,q^2$).

A Hermitian variety U in $PG(2,q^2)$ is called a unital, and is the set of self-conjugate points under a Hermitian polarity π. U contains q^3+1 points and no lines. A line in the plane is either a tangent to U, or a secant which meets U in q+1 points.

Let X be the complex whose vertices are the secants of the unital, and whose top simplices are the points of the unital. There is a complement X^c whose top simplices are the tangents to the unital and whose vertices are the points not on the unital. π sends points of U to tangents of U, and points off U to secants of U ,and so π is an isomorphism between X and X^c.

X^c determines colorings of X as follows. Let p be a point not on the unital. Let $\beta(p)$ consist of the polar line $\pi(p)$ and all the secants passing through p. If we consider all $\beta(p)$, for p on a tangent line and not in U, then we get a coloring. This gives us the usual map $\beta : X^c \to$ B(X). If β were an isomorphism , then we would have an infinite family of self dual complexes.

If q is 2 , then B(X) consists of exactly one coloring. In this case β is onto , but not 1-1. We conjecture that X is self dual, but we can only show

Lemma 1

For q > 2 , $\beta : X^c \to$ B(X) is 1-1.

Proof:

Suppose that $p \notin$U. There are q^2-q secants through p , so if q is greater than 2 , the number of secants through p is greater than 2. In the configuration of points and lines of $\beta(p)$, only p has more than two lines through it , and so β is 1-1. □

We now turn our attention to the Hermitian variety in $PG(3,q^2)$. Let π be a Hermitian polarity and let H be the set of self-conjugate points of π. We recall some basic facts about H. H contains $(q^3+1)(q^2+1)$ points and $(q^3+1)(q+1)$ lines. Through any point of H there are q+1 lines of H , and they constitute the entire intersection of H with the tangent plane at that point. Every plane containing a line of H is a tangent plane of H. The intersection of H wit a non-tangent plane is a unital in that plane.

Let X be the simplicial complex whose vertices are the points of of H and whose top simplices are the lines of H. Here is a simple construction for one type of vertex of B(X). If P is a non-tangent plane, then $P \cap H$ is a vertex of B(X). Indeed, if two points of $P \cap H$ were on a line of H, then the entire line would be in H , and so P would be a tangent plane. However , there are many other kinds of vertices than these .

Suppose that we have a set of lines $L_1, \cdots L_n$ with underlying point set L. If $L \cap H$ is a vertex of B(X) , then we say that the set of lines is a vertex of B(X). Our next lemma shows how to construct many vertices of B(X).

Lemma 2

Suppose that the set $\{L_1, \cdots L_n\}$ of lines is a vertex of B(X) where n $= q^2-q+1$. Then the set $\{\pi(L_1), L_2, \cdots L_n\}$ is also a vertex of B(X) .

Proof:

Every line of H has exactly one point of a vertex of B(X) , and every point of H has q+1 lines of H through it, so the number of points in a vertex of B(X) is $(q+1)(q^3+1)/(q+1) = q^3+1$. Since $q^3+1 = (q+1)(q^2-q+1)$, all the lines L_i are disjoint.

Suppose S is a set of at most q^3+1 points which meets every line of H. There are at most $(q+1)(q^3+1)$ pairs (p,G), where p is a point of S which lies on the line G of H. But the number of pairs is at least $(q+1)(q^3+1)$, for that is the number of lines in H. Thus no line of H contains more than one point of S, so S is a potential vertex of B(X).

Next, L_i is contained in a non-tangent plane whose intersection with H is a unital. Since a secant meets a unital in q+1 points, L_i meets H in q+1 points. Thus, the number of points in $V \cap H$, where V is the underlying point set of $\{\pi(L_1), L_2, \cdots L_n\}$, is at most $(q+1)(q^2-q+1) = q^3+1$.

It now suffices to show that the new set of lines meets every line of H. Choose a line G of H. If it does not meet $L_2, \ldots L_n$, then it meets L_1. Since G lies in H, $\pi(G) = G$, and since G meets L_1, $G = \pi(G)$ meets $\pi(L_1)$. □

Although we will show that X has colorings, there is no obvious complement. If there were a line of $PG(3,q^3)$ disjoint from H, then we could take all the vertices determined by the intersection of H with planes passing through this line. However, there are no exterior lines to H. If L were such a line, let P be a plane containing L. If P is a tangent plane, then $P \cap H$ contains a line, and so H intersects L. If P is not a tangent plane, then $P \cap H$ is a unital, and L is an exterior line to a unital. This is not possible, as unitals do not have exterior lines.

To construct a coloring, let P be a non-tangent plane. $P \cap H$ is a unital, and so has a coloring. Let p be a point on this unital, and let F_i, $1 \leqslant i \leqslant q^2$, be the secants through a point P_i on the tangent line at p. F_i contains q^2-q+1 lines each containing q+1 points of the unital. Define πF_i to be the set of lines πL, where L is a line of F_i. By repeated applications of the last lemma, πF_i is a vertex of B(X).

Theorem 3

$\pi F_1, \cdots, \pi F_{q^2}$ and $P \cap H$ is a coloring of X.

Proof:

Each of the vertices contains q^3+1 points, and there are q^2+1 vertices . Since H contains $(q^2+1)(q^3+1)$ points , it suffices to show that these vertices are all distinct.

πF_i and πF_j do not intersect in H. Suppose that there are lines G_i *in* F_i and G_j *in* F_j such that πG_i meets πG_j in a point p of H. Applying π , G_i and G_j are contained in the plane πp. However , G_i and G_j also lie in the plane P, so P equals πp. This is a contradiction , for P is not a tangent plane.

We claim that if L is not contained in H, then $L \cap \pi L \cap H = \emptyset$. Suppose that $p \in L \cap \pi L \cap H$. $\pi p \cap H$ consists of q+1 lines passing through p, and L meets H in q+1 points , so L cannot pass through p.

$P \cap H$ and πF_i are disjoint. Let $p \in \pi L \cap P \cap H$, where L is a line of F_i . Since F_i is a vertex of a coloring of the unital in P, p lies in a line G of F_i. By Lemma 2 , $(\pi F_i - \pi G) \cup G$ is also a vertex. If L differs from G , then this vertex contains both G and πL. Since p lies in both of them , we have L equals G. The previous paragraph showed that πL and G are disjoint in H, and so cannot contain p. □

Graphs

1. Introduction

Suppose that G is a regular graph of degree r. L(G), the line graph of G, is the simplicial complex whose vertices are the edges of G, and whose top simplices are in 1-1 correspondence with the vertices of G. Given a vertex p of G, the corresponding top simplex consists of all edges which meet p. L(G) is a pure (r-1)-complex. If we ignore the simplicial structure, and consider L(G) as a graph, then this is the usual definition.

Coloring L(G) in r colors is equivalent to coloring the edges of G in r colors such that at every vertex, there are r different colors. In this way we incorporate edge coloring into our general framework.

After reviewing some known resuts on edge coloring, we prove that the line graphs of various graphs constructed from the complete graph K_n and compete bipartite graph $K_{n,n}$ are reflexive, including $K_{2n}\#K_{n,n}$, $K_{n,n}\#K_{m,m}$, and $K_{2n}\#K_{2m}$. We generalize to hypergraphs, and show that the line complex of the complete r-partite hypergraph is reflexive. Next, we study graphs which are of the form Aut(X) for some graph X. We determine all such 3-regular graphs, and find their colorings.

Some Known Results

Many of the results in the literature about coloring L(G) assert the existence (or non-existence) of colorings of a specific graph. In this section we briefly review some of these results.

1 If G has an odd number of edges, then BL(G) $=\emptyset$. This is a trivial result.

2 $B(\hat{L}(G)) \neq \emptyset$ for any G. This follows from Vizing's theorem [Vizing, 1964] To see this, note that if G is r-regular, then $\hat{L}(G)$ uses r+1 colors. These colors can be viewed as being used on the edges of G along with the vertices of G. The color left over by the edge coloring of the edges at a vertex - the color of the vertex - is the

color of the hatted vertex corresponding to the vertex. Vizing's theorem asserts that every r-regular graph has an r+1 coloring, so this gives a coloring of $\hat{L}(G)$.

3 If L(G) has an even number of edges, then BL(G) non-empty implies that BLL(G) is non-empty. [Jaeger, 1974] This is a surprising result, and the proof does not appear to be functorial.

4 BL(G) non-empty implies BL(G#H) non-empty. [Himelwright and Williamson ,1974]

5 If G and H have 1-factors then BL(G#H) is non-empty. [Kotzig] A 1-factor is a set of edges such that every vertex meets exactly one of the edges. A vertex of BL(G) is a 1-factor of G, but not necessarily conversely.

6 BL(G) non-empty implies BL($G \circ H$), BL($H \circ G$), and BL($G \circ 2n\Delta^0$) are non empty. [Wallis] There are also some weaker conditions which give these conclusions.

7 The following special families of graphs have also been studied:

a) K_n, the complete graph on n vertices. There are colorings of L(K_n) for n even ; none for n odd.

b) $K_{n,n}$: it is easy to see that $L(K_{n,n}) \cong \Delta_n \# \Delta_n$ and consequently $BL(K_{n,n})$ is non-empty.

c) $K_{n,n,\ldots n}$ has colorings when rn is even, where r is the number of factors.

d) $B(L(L(K_n)))$ is non-zero exactly when $n \equiv 0$ or 1 mod 4. [Alspach, 1982]

e) The line graph of $P_m \circ nK_1$ has colorings for mn even.

f) The line graphs of the generalized Peterson graphs have colorings ; except of course for the Peterson graph whose line graph has no colorings. [Castagna and Prins, 1972]

2. Reflexive Line Graphs

In this section we prove that certain line graphs built out of $K_{n,n}$ and K_n are reflexive.

Theorem 1

The following graphs have reflexive line graphs:

a) K_{2n}

b) $K_{n,n}$

c) $\hat{K}_{2n}$

d) $\hat{K}_{n,n}$

e) $(K_{2n})\#(K_{2m})$

f) $(K_{n,n})\#(K_{2m})$

g) $(K_{n,n})\#(K_{m,m})$

h) $(P_{2n})\#(K_2)$

i) $K_{2n} {}^{*}K_{m,m}$

j) $K_{n,n} {}^{*}K_{m,m}$

Theorem 2

a) $B^2(L(\hat{K}_{2n+1})) \cong L(K_{2n+2})$

b) $B^2(L(P_{2m+1}\#K_2)) \cong (\hat{P}_{2n+1})$

c) $B^2(L(K_{2m+1}\#K_2)) \cong L(K_{2m+1}\#K_2) \cup$ one more simplex, for $2m+1 \geqslant 5$

Proof:

a) and b) follow from Theorems 3 and 4 [Fisk, 80] ; c) and d) follow from Theorem 1.4.1. e), f), g), i), and j) will follow from a more general result (Theorem 3). We now prove h) along with Theorem 2.b by calculating $B(L(P_m\#K_2))$. $P_m\#K_2$ consists of two circles with m vertices each, joined by spokes. Consider the mapping $\pi: L(P_m\#K_2) \to \hat{L}\ (P_m) \cong \hat{P}_m$ given by sending an inside edge to the corresponding outside edge, and the spokes to the hatted points. A 3-coloring of the edges of $L(P_m\#K_2)$ determines a 3-coloring of the edges of its bounding circles. If m is odd, all 3 colors must be used on either bounding circle, and so the coloring of one circle determines the identical coloring on the other circle. Consequently, $B(\pi)$ is onto; since it is easily seen to be 1-1, it is an isomorphism. This proves 2.b, since $B^2(L(P_m\#K_2)) \cong B^2(\hat{P}_m) \cong \hat{P}_m$.

If m is even, then there is one coloring which is not in the image of $B(\pi)$. Consequently, $B(L(P_m\#K_2)) \cong B(\hat{P}_m) \cup \Delta_2$ where the new triangle contains exactly one point of $B(\hat{P}_m)$, namely the 1-factor of all the spokes. Now $B(B(\hat{P}_m)\cup\Delta_2)$ consists of two copies of $B^2(\hat{P}_m)$ joined along the vertices of the common point, These vertices are all the hatted points of $\hat{P}_m$. Since $\hat{P}_m \cup \hat{P}_m$ is isomorphic to $L(P_m\#K_2)$, we are done. □

Proof:

We begin with a). Consider a coloring f of $L(\hat{K}_{2n+1})$ to be a coloring of the edges and vertices of K_{2n+1} in 2n+1 colors. Since there are an odd number of vertices, every color colors at least one vertex. There are as many colors as vertices, so each colors appears exactly once. Define a map

$$\pi: L(\hat{K}_{2n+1}) \rightarrow L(K_{2n+2})$$

by imbedding $L(K_{2n+1})$ in $L(K_{2n+2})$ and sending all the hatted points to the remaining vertex. By the above, every coloring of $L(\hat{K}_{2n+1})$ is induced by a coloring of $L(K_{2n+2})$, so $B(\pi)$ is onto. $B(\pi)$ is obviously 1-1, so we have an isomorphism. Consequently,

$$B^2(L(\hat{K}_{2n+1})) \cong B(B(L(K_{2n+2}))) \cong L(K_{2n+2})$$

We now begin with the proof of c) which is more interesting. Let n=2m+1. We shall show that the additional simplex has as vertices 1-factors which correspond to the edges joining the two copies of K_n. More precisely, $K_n\#K_2$ consists of two copies of K_n with the corresponding vertices of K_n joined by edges. Let S be the set of the points in $L(K_n\#K_2)$, corresponding to these edges. Let f be a coloring of $L(K_n\#K_2)$. f restricted to either copy of $L(K_n)$ is an n-coloring. The set of all edges in $L(K_n)$ of a given color meets an even number of vertices and since n is odd, for every color there is at least one vertex colored that color by f. There are n colors and n vertices, so every color colors exactly one vertex. Now consider the set of 1-factors $\{\Phi(s) \mid s \in S\}$, where $\Phi: L(K_n \# K_2) \to B^2(L(K_n \# K_2))$ is the natural map. This is the extra simplex of the theorem.

Fix a coloring F of $B(L(K_n \# K_2))$. We begin by determining F on some subcomplexes of $B(L(K_n \# K_2))$. Let K_2 have vertices a and b, and let the two copies of K_n in $K_n \# K_2$ be Δ_a and Δ_b. Choose an n-coloring α of $L(\Delta_a)$ and let γ be a vertex of α. We consider γ as using exactly one edge of S. Given any n-coloring β of $L(\Delta_b)$, there is exactly one vertex γ' of β such that $\gamma \cup \gamma'$ is a 1-factor of $L(K_n \# K_2)$. Consequently, for every $\gamma' \in B(L(\Delta_b))$ a unique vertex γ of α such that $\gamma \cup \gamma'$ is a vertex of $B(L(K_n \# K_2))$. These vertices determine a subcomplex of $B(L(K_n \# K_2))$ which is isomorphic to $B(L(\hat{\Delta}_b))$. By part a), F restricted to this complex is either determined by a vertex p of Δ_b or by S, in the sense that all 1-factors containing a fixed edge through p (or in S in the other case) are colored alike by F.

Choose a different coloring α' of $L(\Delta_a)$ that shares a vertex t with α. Suppose α determines a point p in Δ_a and α' determines $p' \in \Delta_a$. If $p \neq p'$ then we can find edges e_1, e_2, e_3 such that p is not a member of e_2 *or* e_3, $p \in e_1$, $p' \in e_2$ and e_3 , $e_2 \neq e_3$ and there exist 1-factors containing e_1 and e_2, and 1-factors containing e_1 and e_3. Here we assume that n is at least 5. If we add t to these 1-factors, we see that they are colored alike by F (since they contain p) and are also colored differently, since they have two different edges at p'. Consequently, $p = p'$. If either p or p' happens to belong to an edge of t, a similar construction shows that $p = p'$.

We also can not have p determined by α and S determined by α' for similar reasons. Simply choose 1-factors which contain the same edge at p and different ones at S.

We now conclude that all 1-factors containing a fixed edge at $p \in \Delta_a$ (or in S) and a fixed 1-factor of $L(\Delta_b)$ are colored alike by F. Similarly, all 1-factors containing a fixed edge at $p' \in \Delta_b$ (or in S) and a fixed 1-factor of $L(\Delta_a)$ are colored alike by F. Consequently, if we choose a fixed edge at p (or in S) and a fixed one at p' (or in S), then we see that F colors alike all the 1-factors containing these two edges. Let $F(e_a,e_b)$ denote the color assigned to all 1-factors containing e_a and e_b. All 1-factors containing an edge of S are colored alike.

We have three cases. (1) Suppose that S determines F in both Δ_a and Δ_b. Then this is the 'extra' simplex. (2) Suppose that $p \in \Delta_a$ and S determines the 1-factor in Δ_b. Since all 1-factors containing an edge of S are colored alike, this case does not arise. (3) Suppose that $p \in \Delta_a$ and $p' \in \Delta_b$. There are n edges e through p in $L(\hat{\Delta}_a)$ and n edges e' meeting p' in $L(\hat{\Delta}_b)$ giving n^2 pairs (e,e'). If e_1,e_2,e'_1,e'_2 are all distinct, with $p \in e_1,e_2$ and $p' \in e'_1,e'_2$ then there are colorings which determine 1-factors containing e_1 and e'_1, and 1-factors containing e_2 and e'_2, hence $F(e_1,e'_1) \neq F(e_2,e'_2)$. Now construct a graph on the $(n-1)^2$ vertices (e,e'), where e is an edge at p and e' is an edge at P', and neither is in S. Two vertices of this graph are adjacent if the first two coordinates are distinct, and the second two coordinates are distinct. We then have all four distinct, for none of them are in S. F determines a coloring $\tilde{F}$ of this graph by $\tilde{F}(e,e') = F(e,e')$. This graph is exactly $\Delta_{n-1} \# \Delta_{n-1}$ and so all colorings are projections. By symmetry, we may assume that $F(e,e') = F(e,e'') = G(e)$ for any e',e''. This is exactly the statement that F is $\Phi(e)$. □

Theorem 3

Suppose that X is regular of degree r and Y is regular of degree s and each is the disjoint union of complete graphs with an even number of vertices, and complete bipartite graphs. Furthermore, assume that V(X)=V(Y) and E(X) and E(Y) have no edges in common. For any two distinct vertices u and v of V(X) ,assume there is a 1-factor α of X and β of Y such that $\alpha \cup \beta$ contains a cycle that contains u and not v.

Then $L(X \cup Y)$ is reflexive.

Proof:

In a complete graph or a complete bipartite graph, any partial 1-factor is contained in a 1-factor, and every 1-factor is part of a coloring. The same therefore holds true for X and Y. The next Theorem will show that if every partial 1-factor is indeed contained in a 1-factor, then the graph is a disjoint union of complete graphs and complete bipartite graphs.

If f is a coloring of L(X) and g is a coloring of L(Y), then we get a coloring of $L(X \cup Y)$ by using f to color the edges of X and g for the edges of Y. Similarly, if α is a 1-factor of X and β is one of Y, then α and β are 1-factors of $X \cup Y$. Since any choice of α is compatible with any choice of β, we get an embedding of $BL(X) * BL(Y)$ in $B(L(X \cup Y))$. Fix a coloring F of $B(L(X \cup Y))$. We will prove the theorem by showing that F of a 1-factor is the edge of the 1-factor meeting some fixed point p, where we identify the edges at p with the different colors.

Since $B(BL(X) * BL(Y)) \cong B^2L(X) * B^2L(Y) \cong L(X) * L(Y)$, $F| BL(X) * BL(Y)$ has a simple representation. Every top simplex of L(X)*L(Y) is composed of a pair (top simplex of L(X),top simplex of L(Y)), so we can think of $F| BL(X) * BL(Y)$ as a pair (x,y), where $x \in V(X)$, $y \in V(Y)$. If $\alpha \in BL(X), \beta \in BL(Y)$, then $F(\alpha)$ (resp. $F(\beta)$) is the unique edge of α (resp. β) which meets x (resp. y).

Suppose $\alpha \in B(L(X \cup Y))$. We construct a complement $\tilde{\alpha}$ to α as follows: $\alpha| L(X)$ is a partial 1-factor of X, and so is contained in a 1-factor α_1 of X. Similarly, $\alpha| L(Y)$ is contained in a 1-factor β_1 of Y. Define $\tilde{\alpha} = (\alpha_1 - \alpha) \cup (\beta_1 - \alpha)$. Since $\tilde{\alpha} \cup \alpha$ is the union of the two 1-factors α_1 and β_1, and α is a 1-factor, it follows that $\tilde{\alpha}$ is a 1-factor. By hypothesis, these two 1-factors are contained in colorings f and g, whose 1-factors we take to be $\alpha_1, \cdots \alpha_r, \beta_1, \cdots \beta_s$. Now F is known on $\alpha_2, \cdots \alpha_r, \cdots \beta_2, \cdots \beta_s$: $F(\alpha_i)$ is the edge of X which contains x, for $2 \leqslant i \leqslant r$ and $F(\beta_j)$ is the edge of Y containing y for $2 \leqslant j \leqslant s$. There are only two colors left : one edge in E(X) at x and one in E(Y) at y. Letting P denote the edges in E(X) at x or in E(Y) at y, then $\{F(\alpha), F(\tilde{\alpha})\} = \{$ the edges of $\alpha \cup \tilde{\alpha}$ in P $\}$.

Let α be a 1-factor of $X \cup Y$ and $\alpha_z \cup \alpha | X$ a 1-factor of X. We claim α_z can not consist of exactly one edge. For if α_z is a single edge e, then α meets all but two vertices with edges of X. The remaining two vertices must be joined by an edge of Y, which must also be e. Thus e is in $E(X) \cap E(Y)$ - contradiction.

Choose a 1-factor α of $X \cup Y$ such that each of $\alpha | Y$, $\alpha | X$ is lacking at least two edges to be a 1-factor. We can construct four complements to α by combining the two disjoint 1-factors of one of them with the two of the other. For each of the four pairs (α, complement) we get a set of possible edges for their coloring, and $F(\alpha)$ is in the intersection of all of them. But their intersection is just the set of all edges of α in P, so $F(\alpha) \in \alpha \cap P$.

Now choose any coloring of $L(X \cup Y)$ with 1-factors $\{f_i\}$. Every f_i meets P, since $F(\alpha)$ is defined for all α. Since P has r+s edges and there are r+s 1-factors, it follows that each 1-factor meets P in exactly one edge, and so $F(\alpha) = \alpha \cap P$.

By assumption, there are 1-factors α of X and β of Y and an $\alpha\beta$ cycle C such that $x \in C$, $y \notin C$. The new 1-factor $(\alpha - C) \cup (\beta \cap C)$ does not meet P. This is impossible unless x=y, and so we see that $F(\alpha)$ is the edge of α at x and so $L(X \cup Y)$ is reflexive.

□

The proof of this last theorem depended heavily on the existence of a complement. We next show how restrictive this assumption is.

Theorem 4

Suppose that G is a connected graph such that every partial 1-factor is contained in a 1-factor. Then G is K_{2n} or $K_{n,n}$.

Proof:

Suppose G has a subgraph with edges (a,b),(a,c),(b,c),(a,d). The set {(a,d),(b,c)} is a partial 1-factor, and so there is another partial 1-factor F such that the union of the two is a 1-factor. Consider the partial 1-factor $F \bigcup (a,b)$. Since this is contained in a 1-factor, and there are only two vertices not in this union, it follows that (c,d) is an edge. Similarly,(b,d) is an edge. If there is a point adjacent to any of a,b,c,d then the same argument shows that it is adjacent to all of them. Continuing, we see that G is a complete graph, which must be K_{2n}.

Next, suppose that G has an odd cycle with edges $(a_1,a_2),(a_2,a_3), \cdots (a_{2n+1},a_{1})$. Since a odd cycle has no 1-factors, there is another edge meeting this cycle, say (a_1,b). Consider the partial 1-factor $(a_1,b),(a_2,a_3),...(a_{2n},a_{2n+1})$. There is a partial 1-factor F whose union with these edges gives a 1-factor. Considering the partial 1-factor $F \bigcup \{(a_1,a_{2n+1}),(a_{2n},a_{2n-1}), (a_4,a_3)\}$, we see that the two vertices that do not occur, b and a_2, must be adjacent. This gives us a triangle with vertices b,a_1,a_2, and so G is K_{2m} for some m.

We now assume that G has no odd cycles, and hence is bipartite. Let A and B be the vertices of the largest $K_{n,n}$ contained in G. Assume p is a vertex adjacent to q in A, with p not in B. If there is no such p, then by symmetry G is $K_{n,n}$. By taking a partial 1-factor joining A to B, we see p is adjacent to r, for some r in neither A nor B. Choosing any point s of B, and considering the edges (r,p) and (q,s), we see we must have edge (r,s). As in the last paragraphs, p is adjacent to all vertices of A, and r is adjacent to all of B, so $A \bigcup B \bigcup \{p,s\}$ is a larger complete bipartite graph. □

$L(K_6)$ is self dual. Figure 1 labels each edge of K_6 with a 1-factor of K_6. In [Cameron and van Lint, 1980], they use the fact that $L(K_6)$ is self dual to construct outer automorphisms of Sym(6), to construct the Moore graph of degree 6, and to construct the 5-(12,6,1) Steiner system.

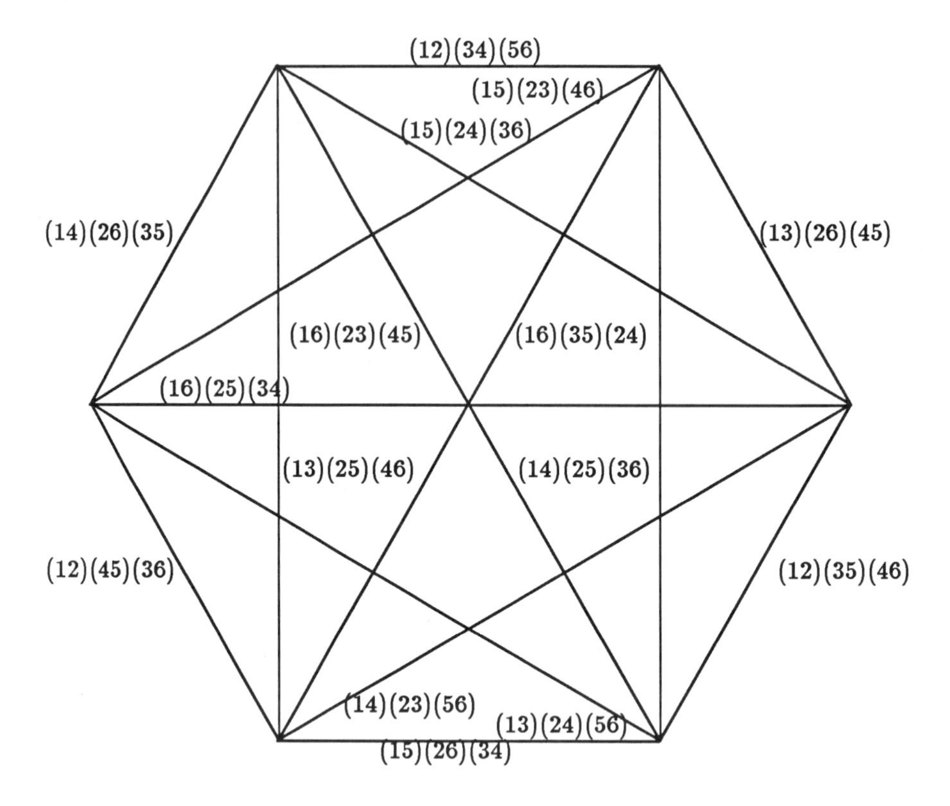

Figure 1

3. Generalized Line Graphs

In this section we indicate how some of the results of the last section may be generalized. We are able to prove a few results, but there are some good open questions.

Suppose that X is a simplicial complex. Define $L_r(X)$, the generalized line graph, to have as vertices the (r-1) simplices of X. Two vertices are adjacent if they intersect. We can also define the generalized line graph for hypergraphs, whose definition we now recall. A hypergraph is a set of vertices along with a collection of subsets of these vertices. A hypergraph differs from a simplicial complex in that not all subsets of a member of a hypergraph are members of the hypergraph. $L_r(X)$ is well defined for hypergraphs.

If p is any vertex of X, then $V_p = \{y \in L_r(X) \mid p \in y\}$ is a set of mutually adjacent vertices of $L_r(X)$. To color $L_r(X)$, we need as many colors as the size of the largest V_p. We call an independent set of $L_r(X)$ a partial 1-factor of X, and an independent set which contains every vertex is called a 1-factor of X. Equivalently, a 1-factor is a set of disjoint r-sets which partitions V(X).

There are two important classes of hypergraphs. The complete r-uniform hypergraph, $K^r{}_m$, has a vertex set with m elements. Its edges are all the r subsets of these m elements. The complete r-partite hypergraph, $K^r{}_{n,n\ldots n}$, has a vertex set which is the union of r disjoint n-sets ; its edges are all those r-tuples which meet each n-set in precisely one vertex. A necessary condition for the existence of a 1-factor of $K^r{}_m$ is that r divides m.

Lemma 1

Every partial 1-factor of either $K^r{}_{rn}$ or $K^r{}_{n,n\ldots n}$ is contained in a 1-factor.

Proof

A partial 1-factor of $K^r{}_{rn}$ consists of t disjoint r-subsets of a set of size nr, while a 1-factor consists of n disjoint r-subsets. If $t<n$ we can find n-r disjoint subsets not meeting the partial 1-factor.

A partial 1-factor of $K^r_{n,n...n}$ is a collection of t r-tuples such that no element appears twice in any coordinate. A 1-factor is a collection of n such r-tuples. Clearly one can find n-t r-tuples to extend the partial 1-factor. □

There are other hypergraphs which satisfy the condition that every partial 1-factor is contained in a 1-factor. For instance, take six vertices a,b,c,d,e,f, then the hypergraph whose edges are (a,b,c),(d,e,f),(a,b,d),(c,e,f) satisfies the condition.

Lemma 2

K^r_{rn} and $K^r_{n,n...n}$ both have the property that $L_r(\bullet)$ has colorings, and every 1-factor is contained in a coloring.

Proof:

The symmetric group Sym(nr) acts on K^r_{rn}, and is transitive on the 1-factors, so it suffices to show that there is some coloring. This was shown by by Baranyai [Baranyai, 1973], but no explicit colorings were constructed.

Similarly, the group Sym(n)$\int$Sym(r) acts on $K^r_{n,n...n}$, and is transitive on 1-factors, so it suffices to construct a coloring. Take each of the r disjoint sets to be a copy of Z_n, the integers mod n, so that a vertex of $L_r(K^r_{n,n...n})$ is an r-tuple $(x_1,x_2, \cdots x_r)$, $x_i \in Z_n$. Let f be any 1-factor of $K^r_{n,n...n}$, and let W be the set of all vertices of $L_r(K^r_{n,n...n})$ which have a 0 in the first coordinate. We claim that the set of 1-factors $\{f+z \mid z \in W\}$ =f+W is a coloring. By $f+z$ we mean $\{\alpha+z \mid \alpha \in f\}$, where α and z are added coordinate by coordinate. Clearly $f+z$ is a 1-factor, so suppose that $f+z \cap f+w \neq \emptyset$. If $\alpha+z=\beta+w$ then since the first coordinates of z and w are 0, and f has exactly one member with any specified coordinate, $\alpha=\beta$ and hence $z=w$. □

The construction shows that n^{r-1} colors are used. We next show that $K^r_{n,n...n}$ satisfies the Helly property [Fisk, 1980].

Theorem 3

Suppose that F is a set of 1-factors of $K^r{}_{n,n...n}$ such that any two 1-factors have non-empty intersection, and every coloring contains exactly one member of F. Then there is a edge of $K^r{}_{n,n...n}$ which is contained in every member of F.

Proof:

We follow the proof of Theorem 4 in [Fisk, 1980]. Choose any 1-factor $f \in F$. If we choose two r-tuples of f, and interchange the values at some coordinate, we get a new 1-factor. All 1-factors of $K^r{}_{n,n...n}$ can be obtained in this fashion, so we can find a pair of 1-factors a and b such that $a \in F$ but $b \notin F$, and they differ by a transposition. Without loss of generality ,we may take them to be (a) and (b) below. We denote the r-tuple $(x,y,y \cdots y)$ by (xy^{r-1}) and $(x,x \cdots x)$ by (x^r).

$\{(1^r),(2^r),(3^r),...(n^r)\}$ a)
$\{(12^{r-1}),(21^{r-1}),(3^r),...(n^r)\}$ b)
$\{1^r),(2n^{r-1}),(32^{r-1}),(43^{r-1}),...(n(n-1)^{r-1})\}$ c)
$\{(1(n-2)^{r-1}),(2(n-1)^{r-1}),(31^{r-1}),(42^{r-1}),...(n-1)(n-3)^{r-1}),(n^r)\}$ d)
$\{(1^r),(2^r),(34^{r-1}),(45^{r-1}),...(n-1)n^{r-1}),(n3^{r-1})\}$ e)
f)
$\{(1^r),(2(n-k+1)^{r-1}),(3(n-k)^{r-1}),...((k-3)n^{r-1}),((k-2)2^{r-1}),((k-1)3^{r-1}),...(n(n-k-2)^{r-1})\}$

The coloring b+W $= \{b+w \mid w \in W\}$ contains some member h = b+g of F. Since $h \cap a$ is non-empty, we can solve for g: if $(1^r) \in (h \cap a)$ then $(1^r)=(12^{r-1})+g$ which implies $g=(0(-1)^{r-1})$; for $(2^r) \in h$ then $g=(01^{r-1})$ and for all other members of (a) we find $g=0$. By symmetry we may assume that $g=(0(-1)^{r-1})$ and so (c) above is in F. Consider the 1-factor (d) above. If we consider the equation $(d+w) \cap a \neq \emptyset$ then we find $w \in \{02^{r-1}),(0^r),(03^{r-1})\}$ while the equation $(d+w) \cap c \neq \emptyset$ leads to $w \in \{(01^{r-1}),(0(-1)^{r-1}),(03^{r-1})\}$ whence $w=(03^{r-1})$ and so $d+(3^{r-1})=(e)$ is in F.

Now consider the family of 1-factors f_k for $2 \leqslant k \leqslant n$. Determining the possible solutions for f_k and (a),(e), and (c), we find that the only solution is that $f_k \in F$, for all n-1 choices of k.

Now suppose that $\alpha \in F$ but (1^r) is not a member of α. α has n members, and at least two of them have 1's in some coordinate. There are (n-2) members of α which could possibly intersect the f_k. However, $f_k \cap f_s = (1^r)$ for any $s \neq k$. Consequently each of these (n-2) members of α intersects a unique f_k. Since there are (n-1) f_k, that completes the proof. □

We next show that $K^r{}_{n,n\ldots n}$ has the extension property. For r = 2, this is the well-known fact that two disjoint permutations are contained in a Latin square.

Theorem 4

Any two disjoint 1-factors of $K^r{}_{n,n\ldots n}$ are contained in a coloring.

Proof:

Suppose that f and g are two disjoint 1-factors. If we consider just the first n-set and the i-th n-set, f and g each determine a permutation of 1...n. Since any two disjoint permutations are contained in a Latin square, we can find n-2 further 1-factors such that they are all disjoint and each n-set (except the first n-set) forms a Latin square. Let H be the set of these 1-factors. Now let Z be the set of all r-tuples whose first two coordinates are 0. We claim that if $z \in Z$ then all z+h are disjoint, for all 1-factors h in H. If $z+h \cap w+k \neq \emptyset$, for $w \in Z$, k in H, then the first two coordinates of h and k agree. However, there is a unique 1-factor in H with any given first two coordinates. It follows h = k and z = w. Consequently, Z+H is a coloring. □

Theorem 5

$L_r(K^r{}_{n,n\ldots n})$ is reflexive.

Proof:

Consider a vertex α of $B(B(L_r(K^r{}_{n,n\ldots n})))$. If any two members of it are disjoint, by the last lemma, they are contained in a coloring and so can not be in α. If they are all disjoint, then by the previous lemma, there is a vertex that they all contain. It follows that $L_r(K^r{}_{n,n\ldots n})$ is reflexive. □

4. Group Graphs

A group graph is a group in the category of graphs and non-degenerate maps, where # is used for multiplication. In other words, there is a map $m{:}G\#G\rightarrow G$ such that

* G is a group under the operation $x \cdot y = m(x,y)$
* m is a non-degenerate map

As an example, AUT(X) is a group graph for any graph X. Conversely, in [Fisk, 1982] we showed that every group graph can be realized as AUT(X) for some graph X. Group graphs also arise as Cayley graphs. The term 'group graph' has been used for Cayley graphs[Teh, 1969], but this usage is not the same as ours. We recall the definition of a Cayley graph:

Given a group G and a set H satisfying (1) the identity is not in H and (2) $H=H^{-1}$ then the Cayley graph X(G:H) has as vertices the elements of G, and as edges (x,xh), for any $x \in G, h \in H$.

Every Cayley graph has left multiplication as an automorphism, but not necessarily right multiplication. If it also has right multiplication as an automorphism, then it is a group graph. The simple condition on H for this to happen is

(3) For all $g \in V(G)$, $gHg^{-1}=H$

Given a group graph, H is defined to be all vertices adjacent to the identity.

5. AUT(G)

In this section we discuss a few facts about automorphisms of group graphs. It is important to remember that AUT(G) consists of automorphisms of the graph and not of the group . The first fact is that as a graph, AUT(G) always contains two copies of G, unless G is abelian in which case there is one copy . Define the maps $r_g : x \rightarrow xg$, $l_g : x \rightarrow gx$. We first claim that r_g gives a copy of the graph G. To see this, let (x,y) be an edge of G. We need to show that (r_x, r_y) is an edge of AUT(G). This is true because if $p \in V(G)$, then (px,py) is an edge of G, since l_p is an automorphism of G. The set of vertices $\{r_x \mid x \in G\}$ has no other edges than these, for if (r_x, r_y) is an edge, then applying r_x and r_y to the identity shows that (x,y) is an edge of G. Next, we show that l_g also gives a copy of G. To show that (l_x, l_y) is an edge of AUT(G), we require that if $p \in V(G)$, (xp,yp) is an edge of G. Let H be such that G = X(G:H). (xp,yp) is an edge of G iff $(xp)^{-1}yp$ is in H. Now since (x,y) is an edge, $x^{-1}y \in H$ and hence $p^{-1}x^{-1}yp \in H$ by the third condition for a group graph. Of course, if G is abelian, then left and right multiplication coincide. There is a converse to this :

Lemma 1

If X = X(G,H) is a Cayley graph and AUT(X)=X, then G is abelian.

Proof:

The vertices of AUT(X) are given by $l_g : x \rightarrow gx$ since AUT(X)=X. Since AUT(X) = X, X is also a group graph, so right multiplication must also be an automorphism. Since all r_g are distinct, for every g there is an h such that $r_g = l_h$. This implies that xg = hx whence G is abelian. □

We now show that if X = X(G:H) is just a Cayley graph, then AUT(X) may not have any edges at all. The maps l_g are indeed automorphisms of X, and hence vertices of AUT(X), but they may not necessarily be adjacent. As we saw above, for (l_x, l_y) to be an edge of AUT(X), we must have $p^{-1}x^{-1}yp \in H$ for all p in V(X). Setting p the identity shows that x and y must be adjacent. However this is such a strong condition that it may never be satisfied. As a concrete example, Godsil [Godsil, 1983] constructs infinitely many non-abelian 3-regular Cayley graphs X = X(G:H) such that the automorphism group of X is G. Now if AUT(X) = X, then X would be a group graph, but then

AUT(X) would be larger than X, since X is non-abelian. Consequently, at best AUT(X) is one or two regular.

Examples

S_n S_n is a group graph and $\mathrm{AUT}(S_n) = 2S_n \# S_n$.

P_n Identify P_n with the integers mod n, and define $P_n \# P_n \to P_n$ by $(x \# y) \to x+y$. This is a group of automorphisms of P_n, so P_n is a group graph. From [Fisk, 1982] $\mathrm{AUT}(P_n) = 2P_n$ if n is not 4 and $\mathrm{AUT}(P_4) = P_4 \# K_2$.

$K_{n,n}$ $K_{n,n}$ is usually not a group graph, but $\mathrm{AUT}(K_{n,n})$ is interesting to compute. Let the two sets of n disjoint points of $K_{n,n}$ be A and B. If α is an automorphism, either $\alpha(A)=A$ *or* $\alpha(A)=B$. In case $\alpha(A)=A$ then $\alpha(B)=B$ and so α is determined by two permutations. There are therefore $(n!)^2$ such α. In case $\alpha(A)=B$ there are again $(n!)^2$ automorphisms. If $\alpha(A)=A$ and $\beta(A)=A$ then α and β are not adjacent., whereas if $\beta(A)=B$ then α and β are adjacent. Consequently, all α with $\alpha(A)=B$ are adjacent to all β with $\beta(A)=B$ and no such α is adjacent to a γ with $\gamma(A)=A$. This shows that $\mathrm{AUT}(K_{n,n}) = K_{(n!)^2,(n!)^2}$.

$K_{n,n,\ldots n}$

We can generalize the previous example to see that

$$AUT(K_{n,n,\ldots n}) = S_r \circ (K_{(n!)^r})$$

where r is the number of n's. Each automorphism of $K_{n,n,\ldots n}$ permutes the r blocks of n points, and any map of one block to another is acceptable. For two maps to be adjacent, all that is needed is that the block permutations be adjacent.

6. The 3-Regular Group Graphs

In this section we determine all the 3-regular group graphs.

Theorem 1

A finite 3-regular group graph is either $P_n \# K_2$ or $P_n \# K_2$ with a twist (see Figure 2).

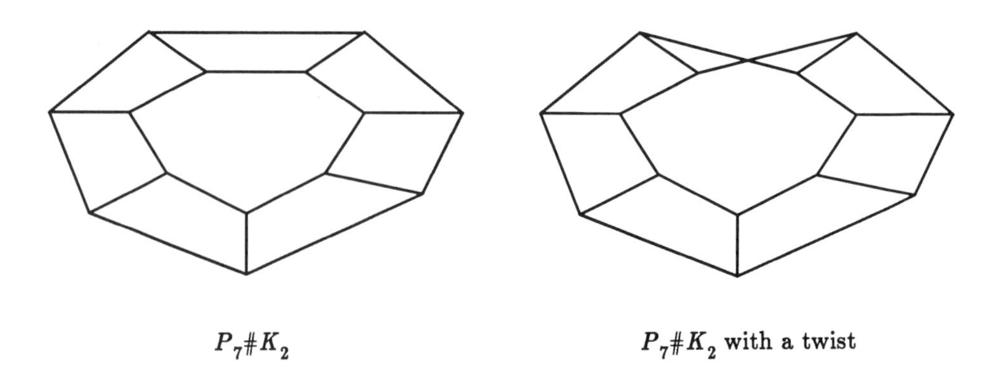

Figure 2

Proof:

If X = X(G:H) is 3-regular, then H has three elements that we may take to be $\{1,2,3\}$. By using the three relations for a group graph, we will determine relations between the elements of H, which will determine G and H.

First, consider H^{-1}. As sets, $\{1,2,3\}^{-1} = \{1^{-1},2^{-1},3^{-1}\}$. Either $i = i^{-1}$ for some i or they are all distinct. In the latter case we have either $1=2^{-1}$, $2=3^{-1}$, $3=1^{-1}$ or $1=3^{-1}$, $2=1^{-1}$, $3=2^{-1}$. In either case all three elements are equal, but 1,2 and 3 are assumed to be distinct. Without loss of generality, we may thus assume that $1=1^{-1}$ or 1^2 is the identity. We have two cases:

$$2^2 = 3^2 = \text{identity} \tag{1a}$$

$$2 = 3^{-1} \tag{1b}$$

Next, consider $1^{-1}H1 = H$. This leads to the two cases:

$$12=21, 13=31 \tag{2a}$$

$$12=31,13=21 \tag{2b}$$

From $2^{-1}H2 = H$ we get the two cases

$$12=21,32=23 \tag{3a}$$

$$12=23,32=21 \tag{3b}$$

Finally, considering $3^{-1}H3=H$ we have

$$13=31,23=32 \tag{4a}$$

$$12=23,32=21 \tag{4b}$$

Of these sixteen combinations of the cases, most do not give rise to a group graph. Let (i,j,k,l) denote the choices 1i,2j,3k,4l and let - denote any choice. First, (-,a,b,-) is not possible. If so, 12 = 21 and 12 = 31 so 2 = 3. By assumption, all elements are distinct. Next, (-,b,a,-) is not possible, for then 31 = 12 = 21 and so 1 = 2. (-,b,-,a) not possible, for 31 = 13 = 21 implies 2 = 3. We are left with four possibilities:

(b,b,b,b)

This is not possible, for 23 = id implies 32 = id and so 23 = id = 32 = 13 implying 1 = 2.

(b,a,a,a)

The group is abelian, with generators 1,2,3 satisfying $1^2=id$, $2=3^{-1}$ .Consequently, we have an infinite family of groups Z_2+Z_r for any r, where r is the order of element 2. The graph is $P_n\#K_2$.

(a,b,b,b)

From the relations we see that all products of 3 terms reduce to one term, and so there are six elements id,1,2,3,12,21 . The graph is seen to be $K_{3,3}$ which is also $P_3\#K_2$ with a twist.

(a,a,a,a)

All generators commute and have order 2, so the group is $Z_2+Z_2+Z_2$ and the graph is the 3- cube, which is $P_4\#K_2$.

We have not found all the 3-regular group graphs yet, for there might be further relations between the generators. The two small graphs have as quotients only K_4 which is $P_2\#K_2$ with a twist. In the remaining case, the generators satisfy $1^2=id=2^r$, $2^{-1}=3$. Any further relation must be of the form $1^i = 2^j$, so there are two possibilities: $1=2^j$ or $id=2^j$. The second case gives $Z_2+Z_{(j,r)}$ which we've already seen. In the other case, there are generators 2^j, 2, 2^{-1} with b^j not equal to the identity. The order of 2 is therefore 2j and the resulting graph is the twisted $P_n\#K_2$. □

We can compute the colorings of all the 3-regular group graphs. Let C_n denote $P_n\#K_2$ and let A_n denote $P_n\#K_2$ with a twist.

Theorem 2

X	B(X)	$B^2(X)$
$L(P_{2n}\#K_2)$	$B(\hat{P}_{2n})$	$\hat{P}_{2n}$
$L(P_{2n+1}\#K_2)$	$B(\hat{P}_{2n+1})+\Delta$	$\Delta\times\hat{P}_{2n+1}$
$L(A_{2n})$	$B(\hat{P}_{2n})$	$L(A_{2n})$
$L(A_{2n+1})$	$B(\hat{P}_{2n+1}) \coprod \Delta$	$\hat{P}_{2n+1}$

Proof:

The results for C_n were proved in Section 2. Define a map $\alpha:L(A_n)\to\hat{P}_n$ as follows. Let A_{2n} have vertices 1,2,...2n and edges $(i,i+1),1\leqslant i=<2n$ and $(i,i+n),1\leqslant i\leqslant n$. Let $\hat{P}_n$ have triangles $(i,i+1,x_i),1\leqslant i\leqslant n$. Let $\alpha(i,i+1)=i$ mod n and $\alpha(i,i+n)=x_{i-1}$. A typical triangle of $L(A_n)$ is $\{(i-1,i),(i,i+1),(i,i+n)\}$ and is mapped by α to $\{(i-1,i,x_{i-1})\}$, and so α is a NDG map. We show that $B\alpha$ is a isomorphism for n even, and for n odd, $B(A_n)=(B\alpha)B(\hat{P}_n) \coprod \Delta$. Given a coloring f of $\hat{P}_n$, the induced coloring $\tilde{f}$ on $\hat{P}_n$ satisfies $\tilde{f}\,(i,i+1)=\tilde{f}\,(i+n,i+n+1)$, since these two edges are sent to the same edge of $\hat{P}_n$. Conversely, if a coloring of $L(A_n)$ has all pairs of edges (i,i+1) and (i+n,i+n+1) colored alike, then the coloring of $L(A_{n})$ is induced by a coloring of $\hat{P}_n$. We thus assume that there is a coloring g and an index i of $L(A_n)$ such that g(i,i+1)=1 , g(i+n,i+n+1)=2, where the colors are 1,2,3. Since (i,i+n) meets both (i,i+1) and (i+n,i+n+1), it is colored 3 by g. Then, (i-1,i) is colored 2; (i-1+,i) is colored 1, and (i-1,i-1+n) is colored 3. Continuing, we see that all (j,j+n) are colored 3, and the remaining edges are colored 1 or 2.

If n is even, then since for any k, $g(i,i+1)=g(i+2k,i+2k+1)$, we see g(i,i+1)=g(i+n,i+n+1). However, these were assumed to be colored 1 and 2, so there is no such g. However, if n is odd, these is such a coloring g and it is unique. We next claim that none of g's 1-factors are in the image of $B\alpha$. Since $n \equiv 1$ mod 2, we see that if the set of edges colored 3 were in the image of $B\alpha$ then all the vertices x_i would be colored alike, but you can not two-color an odd circle. Neither of the two other 1-factors can be in the image of $B\alpha$, for they do not have antipodal edges colored alike. Consequently, we have one triangle which is disjoint from $B\alpha$; this completes the proof.

□

Complexes With a Structure Group

If we consider the results of Chapter 3, we see that much of the complexity comes from that although we are using projective spaces, the maps used are not necessarily projectivities. In this chapter we will allow a 'structure group' to be imposed on the complexes. If the structure group is $PGL(n,q)$, all the maps will be projectivities.

In the first section we transfer basic results from Chapter 1 to complexes with a structure group. We then give a collection of examples with various structure groups. In the third section, we prove a few general results about matrix groups, while in the next section we specialize to PGL(2,q). The main results generalize and simplify Chapter 3. We briefly study PGL(2,q)-colorings of the hyperbolic quadric in PG(3,q), and relate them to PGL(2,q) Latin squares. Next, in an attempt to find a PGL(2,q) analog of $L(K_n)$, we study a certain complex constructed from elliptic involutions in PGL(2,q). Finally, we discuss the extension problem for G-Latin squares, and conclude with a discussion of Latin squares over the Affine group.

1. Introduction

We begin by choosing a subgroup G of Sym(n). We consider Sym(n) and G to act on $\{1,2, \cdots n\}$. We will also assume that G is transitive ; otherwise our theory will be vacuous. If we let Δ be an n-1 simplex with vertices $\{1,2 \cdots n\}$ then we view G as a group of automorphisms of Δ.

Let X be a pure n-1 complex. A *G-structure on X* consists of maps $\{f:\Delta \rightarrow X\}$ where

(1) every n-1 simplex of X is the image of at least one f

(2) if f and g are two maps with the same image, then $g^{-1}f : \Delta \rightarrow \Delta$ is an element of G.

We call the maps f the *structure maps* for X. The definition is similar to the usual definition of differential manifolds. In the rest of this section we will prove analogs of the results of Chapter 1.

If X and Y have G-structures, a non-degenerate map $h : X \to Y$ is a *G-map* if for all structure maps $f : \Delta \to X$ and $g : \Delta \to Y$ such that Im(hf) = Im(g), we have $g^{-1}hf : \Delta \to \Delta$ is in G. Equivalently, all the maps hf are structure maps for Y.

We begin with how to put a G-structure on objects created from objects with a G-structure. Assume X and Y have G-structures. $X \coprod Y$ has the obvious G-structure whose maps are all those of X and of Y. $X \times Y$ has as vertices all (x,y), for $x \in X$ and $y \in Y$. If $f : \Delta \to X$ is a structure map of X and $g : \Delta \to Y$ of Y, then we get a structure map $f \times g : \Delta \to X \times Y$ by setting $f \times g(i) = (f(i), g(i))$, and all structure maps are obtained in this way. For $X \# Y$, the structure maps are of two types: (1) $(f \# q)(i) = f(i) \# q$ and (2) $(p \# g)(i) = p \# g(i)$, where $p \in X$, $q \in Y$.

We define Hom(X,Y;G) to have as vertices all G-maps from X to Y. A map $f : \Delta \to$ Hom(X,Y;G) is a structure map if for all $x \in X$ the map $g : \Delta \to$ Y given by g(i) = f(i)(x) is a structure map for Y. Aut(X;G) consists of all the top simplices of Hom(X,X;G) which consist solely of automorphisms along with the same structure maps.

The vertices of B(X;G) consist of all $f^{-1}(p)$ where $f : X \to \Delta$ is a G-map and $p \in \Delta$. Such maps are called G-colorings. The corresponding structure map $\tilde{f} : \Delta \to B(X;G)$ is given by $\tilde{f}(i) = f^{-1}(i)$.

Theorem 1

Suppose X,Y and Z have G-structures. Then

(a) Hom(X,Y;G) # X $\to$ Y is a G-map.

(b) Hom(X,Y;G) # Hom(Y,Z;G) $\to$ Hom(X,Z;G) is a G-map.

(c) If $f : X \to Y$ is a G-map, so is $Bf : B(Y;G) \to B(X;G)$.

(d) $\Phi : X \to B^2(X;G)$ is a G-map.

(e) Hom(X,B(Y;G);G) $\cong$ Hom(Y,B(X;G);G).

(f) Hom(X,Hom(Y,Z;G);G) $\cong$ Hom(X#Y,Z;G).

Proof:

(a) There are two types of structure maps of the left hand side. Consider first the structure map $f : \Delta \to \mathrm{Hom}(X,Y;G)\#p$ where $p \in X$. The map h of the theorem is the evaluation map, so hf(i) = f(i)(p). This is a structure map for Y by the definition of G-structure in Hom(X,Y;G). Next, let $k : \Delta \to f\#X$ where $f \in \mathrm{Hom}(X,Y;G)$. hk(i) = f(k(i)) which is a structure map for Y since k is one for X and f is a G-map.

(b) The proof is similar to (a) and omitted.

(c) Let $g : \Delta \to B(Y;G)$ be a structure map, so that there is a G-coloring f satisfying $g(i) = f^{-1}(i)$. gf is a G-coloring of X and $Bf(g)(i) = f^{-1}g^{-1}(i) = (gf)^{-1}(i)$ and so Bf(g) is a structure map for B(X;G).

(d) The proof is similar to the above and omitted.

(e) In the S_n case of Chapter 1 we showed that the mapping of maps

$$(f : X \to B(Y)) \quad \to \quad (\Phi(Bf) : X \to B^2(Y) \to B(X))$$

was a bijection and preserved simplices. Since Φ and B preserve G maps, (e) follows.

The proof of (f) is easy and omitted. □

Theorem 2

(a) If G is 2-transitive, X and Y connected, then

$B(X \times Y;G) \cong B(X;G) \coprod B(Y;G)$

(b) $B(X \coprod Y;G) \cong B(X;G) \times B(Y;G)$

Proof:

(b) is obvious. For (a), we see from the proof of the S_n version in [Fisk, 1978] that we only need to prove it for $X = Y = \Delta$. In this case the underlying graph of $\Delta \times \Delta$ is the same for G and S_n. Indeed, if $x \neq z$ and $y \neq w$ there is a g in G such that g(x)=y and g(z)=w. Consequently, (x,y) and (z,w) are in the same top simplex of $\Delta \times \Delta$ for G, and hence joined by an edge. Since $B(\Delta \times \Delta;S_n) \cong \Delta \coprod \Delta$, and the projection maps are in

G, the result follows. □

Note

If the assumption is false, (a) may fail. See Example 6 of the next section.

One of the most interesting complexes which arises is $\mathrm{Aut}(\Delta;G)$. A top simplex is a set of n permutations, each one in G, such that if we read each column, we also get permutations in G. These are called G-Latin squares. Here is a simple result on their existence.

Theorem 3

Let G be a subgroup of S_n. If G contains an element that determines an orbit of length n, then there are G-Latin squares. Equivalently, $\mathrm{Aut}(\Delta;G) \neq \emptyset$.

Proof:

Let $\sigma \in G$ have an orbit of length n. It suffices to construct a G-map $F : \Delta \# \Delta \to \Delta$; if we fix the first coordinate we get the rows of the G-Latin square. Fix $w \in \Delta$ and choose $u,v \in \Delta$. Since there is only one σ orbit, there are unique i and j such that $\sigma^i(w)=u$ and $\sigma^j(w)=v$. Define $F(u,v)=\sigma^{i+j}(w)$. For any simplex of $\Delta \# \Delta$, F is just the action by σ to some power, and is therefore a G-map. □

Note

We have not shown that $B(\Delta \# \Delta;G) \cong \mathrm{Aut}(\Delta;G)$. The difficulty is that the vertices of $B(\Delta \# \Delta;G)$ are permutations which are not necessarily in G. There may be a isomorphism here, for they often agree, but this is a open question.

2. Examples

In this section we will give some examples of complexes with various structure groups.

(1) S_n

In this case there are no restrictions on any maps other that they are non-degenerate, so all the results of Chapter 1 apply.

(2) The trivial group

Suppose that the structure group G is the trivial group. In any G-complex X, every top simplex has a unique structure map from Δ to it. X has at most one coloring f, which necessarily satisfies $f(p)=g^{-1}(p)$ for any structure map g whose image contains p. Any Hom is empty, for there are no adjacent maps.

(3) Alt(n)

Let the structure group G be Alt(n), the group of even permutations. Alt(n) is 2-transitive (n > 3) so Theorem 5.1.2 applies. In [Fisk, 1983] we showed that there are top simplices. We have not been able to show that Alt(n) is reflexive, except that computer calculations have found this to be so for n = 5.

For n = 4, $B(\Delta^3 \# \Delta^3; Alt(4)) \cong 3\Delta^3$. $\Delta^3 \# \Delta^3$ is reflexive, but $B(3\Delta^3; Alt(4)) \cong \Delta^3 \times \Delta^3 \times \Delta^3 \neq$ Alt(4). For n = 5, a computer calculation shows that Y $= B(\Delta^4 \# \Delta^4; Alt(4))$ is a simplicial complex with 60 vertices and 72 top simplices. Y is regular, every point of Y is in 6 tetrahedra, and $B(Y; S_5) \cong B(Y; Alt(5))) \cong \Delta^4 \# \Delta^4$.

For n = 4 we have a nice example of how the choice of structure maps affects the existence of Alt(4) colorings. For any simplex of $\Delta^3 \# \Delta^3$ there are two choices of structure maps, which we think of as odd and even. Suppose that a rows and b columns are given odd structure maps. The choice of rows does not matter, so call the complex L(a,b). We then have

$$B(L(0,0),Alt(4)) \cong 3\Delta^3$$
$$B(L(2,2),Alt(4)) \cong \Delta^3$$
$$B(L(0,4)),Alt(4)) \cong 3\Delta^3$$
$$B(L(4,0);Alt(4)) \cong 3\Delta^3$$

and the rest of the L(a,b) have no colorings.

Finally, an interesting fact about Latin cubes. An Alt(4) Latin cube is an Alt(4) map from $\Delta^3 \# \Delta^3 \# \Delta^3$ to Δ^3. We found that $B(\Delta^3 \# \Delta^3 \# \Delta^3; S_4)$ had 2304 top simplices, but $B(\Delta^3 \# \Delta^3 \# \Delta^3; Alt(4)) \cong 4\Delta^3$.

(4) Alt(2)

Since Alt(2) is the trivial group, an Alt(2) structure map on a graph X is map from Δ to an edge of X. In other words, we choose a direction for each edge; Alt(2) structures correspond to directed graphs.

(5) Alt(4)

Suppose that M is a triangulation of an orientable 2-manifold, and choose structure maps $f : \Delta^3 \to \hat{M}$ such that the maps are orientation preserving. This gives an Alt(4) structure on M which is unique, given the orientation. Let g be an Alt(4) coloring of M. If we have two triangles ABC and ABD of M, then if we know the coloring on ABC, we know the coloring on two of the vertices of the tetrahedron over ABD. Only one of the two possible colorings preserves the orientation, and that is the coloring with g(C) $\neq$ g(D). From [Fisk, 1978], it follows that if M has an Alt(4) coloring, then M is 3-regular. However, if M is 3-regular, M still might not have such a coloring. If there is a coloring, it is unique. Consequently, B(M;Alt(4)) is either $\emptyset$ or Δ^3. These observations have simple generalizations to n-manifolds.

(6) Cyclic group of maximum order

Let G be the cyclic group of order n. We may take a generator σ to be the cycle $(1,2, \cdots n)$. There is exactly one G-Latin square with a given first row, so $G \cong \Delta$ and $B(AUT(\Delta;G);G) \cong G$. G is not 2-transitive. In fact, for this G, $\Delta \times \Delta$ is seen to be $n\Delta$ and so the conclusion of Theorem 5.1.2(b) fails here.

(7) The 1-dimensional Affine Group

If q is a power of a prime, the 1-dimensional affine group AF(1,q) is $\{ x \to ax+b \mid a \neq 0; a,b \in GF(q)\}$. AF(1,q) has a simple simplicial structure. Suppose that f(x) = ax+b and g(x) = cx+d are disjoint permutations. In order that ax+b $\neq$ cd+d for all x, we must have a=c and $b \neq d$. All top simplices are determined by one parameter a, and there are q-1 of them. Consequently,

$$AF(1,q) \cong (q-1)\Delta$$

Since AF(1,q) is 2-transitive, Theorem 5.1.2 applies and so

$$B^2(AF(1,q);AF(1,q)) \cong B^2((q-1)\Delta,AF(1,q)) \cong (q-1)\Delta \cong AF(1,q)$$

Note that AF(1,4) $\cong$ Alt(4), and AF(1,3) $\cong$ S_3. This latter relation 'explains' why $B(\Delta^2 \# \Delta^2) \cong 2\Delta^2$.

(8) PGL(n,q)

PGL(n,q) is the group of projectivities of the projective space PG(n-1,q) which are given by matrices. It is a permutation group on $q^n+q^{n-1}+\cdots q+1$ elements. PGL(n,q) has elements, known as cyclic projectivities [Hirschfeld, 1979], which determine one orbit, so by Theorem 5.1.3 there are PGL(n,q) colorings.

The criterion for disjointness of two members of PGL(n,q) is interesting. Let A and B be two matrices representing elements of PGL(n,q). A and B are disjoint iff for all x, A(x) and B(x) are distinct in PG(n-1,q). Equivalently, there is no $\lambda \in GF(q)$ such that $B^{-1}A(x) = \lambda x$. Thus, A and B are disjoint iff $B^{-1}A$ has no eigenvalues.

Consider PGL(2,q) for small q. If q is 2, PGL(2,2) is all six permutations of the three points of PG(1,2), so PGL(2,2) $\cong S_3$ both as a group and as a complex. PGL(2,3) is all permutations of the four points of PG(2,3), and so $PGL(2,3) \cong S_4$. The group of all collineations of PG(2,4), $P\Gamma L(2,4)$, is isomorphic to Sym(5), so $P\Gamma L(2,4) \cong S_5$. PGL(2,4) consists of all even permutations, so PGL(2,4) $\cong$ Alt(5).

Finally, PGL(2,5) is isomorphic to Sym(5) as a group, but calculations show that PGL(2,5) is not isomorphic to S_5 as a complex. From [Hirschfeld, 1979] we find the isomorphism of PGL(2,5) with Sym(5) is as follows: every element of PGL(2,5) permutes the 5 objects called synthemes and every permutation of synthemes is induced by a unique element of PGL(2,5). If PGL(2,5) and S_5 were isomorphic, then we may assume that the identity of one goes to the identity of the other and so the elements adjacent to the identity in one are sent to elements adjacent to the identity in the other. The map $z \to 2/z$ has no fixed points and is adjacent to the identity. It may be checked that the map does fix a syntheme, and hence there is no isomorphism.

(9) A Combinatorial example

Let $n = \binom{r}{s}$ and take the n vertices of Δ to be all subsets of size s from a set of size r. If we take G to be Sym(r) acting on the subsets of size s, then G acts transitively on Δ, but not 2-transitively.

3. Matrix Groups

We expect that the theory of complexes with PGL(n,q) as structure group should be the q-analog of complexes with structure group S_n. We are, however, not able to prove the basic conjecture : is PGL(n,q) PGL(n,q)-reflexive? PSL(n,q) is the analog of Alt(n), and we know almost nothing about complexes with either of these groups for structure groups. We do know that PGL(n,q) has colorings, a consequence of Singer's Theorem on cyclic projectivities. We do not know if PSL(n,q) has colorings, or even what its dimension is. We will first prove some simple results about the structure of some of these complexes ; the rest of the section will study particular PGL(n,q) complexes which arise from geometric considerations.

Our first theorem gives some local structure of PGL(n,q). Recall that the number of points t(n) in PG(n-1,q) is $q^n + q^{n-1} + \cdots + q + 1$.

Theorem 1

Two top simplices of PGL(n,q) intersect in zero or exactly t(i) points for some i, $0 \leqslant i \leqslant n-1$.

Proof:

A structure map for a top simplex of PGL(n,q) corresponds to a map $F{:}PG(n-1,q) \# PG(n-1,q) \to PG(n-1,q)$. The 'rows' $F(p,\bullet)$ are the vertices of the top simplex. It suffices to show that if $F(p,\bullet)$ is known for a set of vertices $\{p_1,\ldots p_r\}$ then $F(x,\bullet)$ is uniquely determined for all x in the subspace W spanned by the $\{p_i\}$. Since $F(\bullet,y)$ is a projectivity for any choice of y, there is a matrix M_y such that $F(p_i,y) = M_y \bullet p_i$. From this representation we see that F(x,y) is uniquely determined for all $x \in W$. □

Corollary 2

Any two distinct simplices of PGL(2,q) meet in at most 2 points.

Our state of knowledge of GL(n,q), the group of all invertible n by n matrices over GF(q), is also slim. And, we know nothing about SL(n,q) at all.

Theorem 2

GL(n,q) is non-empty. Equivalently, there are GL(n,q) Latin squares.

Proof:

The proof is similar to the PGL(n,q) case. In place of a cyclic projectivity we need a cyclic element of GL(n,q). That such exist follows from a similar construction, using a primative root instead of a sub-primative root. □

Theorem 3

Any two intersecting top simplices of GL(n,q) meet in q^i points for $0 \leqslant i \leqslant n$.

Proof:

Similar to Theorem 1 and omitted. □

In GL(n,q), two matrices A and B have disjoint transformations iff AB^{-1} does not have 1 as an eigenvalue. There is a projection $\pi: GL(n,q) \to PGL(n,q)$ but this is not a non-degenerate map since the dimension of the left hand side is greater than that of the right hand side. If AB^{-1} has eigenvalue $\lambda \neq 1$, then A and B are disjoint in GL(n,q) but $\pi(A)$ and $\pi(B)$ are not disjoint in PGL(n,q). However, if $\pi(A)$ and $\pi(B)$ are disjoint, then A and B are too.

There are two general questions that we can ask about arbitrary matrix groups.

(1) Given a set of non-singular matrices over a field F what is the structure of a maximal set S of elements such that for any two $x,y \in S$, xy^{-1} has no eigenvalues?

(2) If we give the above set of matrices an adjacency structure defined by making (x,y) an edge iff xy^{-1} has no eigenvalue, what can be said about its graph-theoretic structure? In particular, what is its chromatic number?

In addition to the groups P Γ L(n,q), PGL(n,q), PSL(n,q), GL(n,q), SL(n,q), we can also consider O(n,q) or orthogonal matrices. In the cases where the matrices do not act projectively, the criterion for adjacency is that 1 is not an eigenvalue.

4. Colorings of PGL-structures

In this section we investigate the colorings of various PGL(n,q) structures.

Proposition 1

There is a natural PGL(n,q) structure on any set of n-1 flats of PG(m,q) for any $m \geqslant n-1$.

Proof:

Fix some (n-1)-flat F of PG(m,q) and identify it with PG(n-1,q). Given any (n-1)-flat G of PG(m,q), there are perspectivities which carry G to F. These are the structure maps of G. □

This is the not only possible structure we will ever consider for subsets of flats of projective spaces. Consider the set of all lines in PG(3,q). Take these lines as the vertices of a simplicial complex. For each point of PG(3,q), we make a top simplex consisting of all the lines through that point. Since the set of all lines through a point is isomorphic to PG(2,q), this complex has a PGL(3,q) structure.

Let S be a collection of lines in PG(2,q), and give S the incidence structure of the theorem. Let V be the set of points of PG(2,q) which lie on some line of S. We define the complement S^c to S as follows. The top simplices are all points of PG(2,q) not in V. The 'vertices' of S^c are all lines which meet some point not in V. A structure map is defined by sending all lines through a point in S^c to their intersection with another line L, and identifying L with PG(1,q). Any choice of L gives rise to the same PGL(2,q)-structure. Equivalently, if π is a polarity which sends points of PG(2,q) to lines and vice versa, then S^c is equivalent to $\{\pi(p) \mid p \notin S\}$.

S^c has a complement S^{cc}, whose lines are all those lines missing points of S^c. S is generally isomorphic to S^{cc}. We can use S^c to get a $PGL(2,q)$-coloring of S. Suppose that $p \in PG(2,q) - V$. We construct a $PGL(2,q)$ coloring of S as follows. Choose a line L in S. It suffices to define a map from V to L, for we can go to any other choice of PG(1,q) by a projectivity. If L' is any line of S, define β from L' to L as the perspectivity from p. This gives us a map $\beta : S^c \to B(S;PGL(2,q))$. β is not always an isomorphism ; in the remainder of this section we will study B(S;$PGL(2,q)$) in detail.

We first give a geometric description for colorings of S. Let f be a map $f: S \to PG(1,q)$ whose restriction to any line of S is a projectivity. Given any two lines L_i and L_j of S, we get a projectivity f_{ij} from L_i to L_j by sending L_i to PG(1,q) by f, and going back to L_j by f^{-1}. The point $L_i \cap L_j$ is fixed by f and any projectivity fixing a point is a perspectivity [Hartshorn ,1967]. Consequently, there is a point p_{ij} such that f_{ij} is the projection from p_{ij}. p_{ij} is not on either L_i or L_j. The colorings of S in the image of β are those for which all the p_{ij} are all equal. Define L_{ij} as the intersection of L_i and L_j. Our first result is that the p_{ij} determine colorings.

Theorem 2

Let $\{L_1, \cdots, L_k\}$ be a set of lines in PG(2,q). Let $\{p_{ij} \mid 1 \leqslant i < j \leqslant k\}$ be a set of points such that

(1) $p_{ij} \notin L_i \cup L_j$

(2) If L_a, L_b, L_c are not concurrent then p_{ab}, p_{bc}, L_c are colinear

(3) If L_a, L_b, L_c are concurrent, then $f_{ab} f_{bc} = f_{ac}$

then the p_{ij} arise from a unique PGL(2,q) coloring.

Proof:

It suffices to show that $f_{1a} f_{ab} f_{bc} \cdots f_{\alpha 1}$ is the identity. This will follow if we can show for all lines L_a, L_b, L_c that $f_{ab} f_{bc} = f_{ac}$. If the lines are concurrent, then this is assumption (3). If they are not concurrent, then $f_{ac}(L_{ac}) = f_{ab} f_{bc}(L_{ac}) = L_{ac}$, $f_{ac}(L_{bc}) = f_{ab} f_{bc}(L_{bc}) = (p_{ac} L_{bc}) \cap L_a$, and $f_{ac}(p_{ac} p_{bc} \cap L_c) = f_{ab} f_{bc}(p_{ac} p_{bc} \cap L_c) = L_{ab}$. Consequently, the two maps agree at three points, and so are equal by the Fundamental Theorem of Projective Geometry. □

The third assumption could be replaced by a more geometric one, but we will be able to use (3) because we will only choose a generating set of f_{ab} and define the other maps using these projectivities.

We proceed to determine some properties of the p_{ij}.

(1) If L_a, L_b, L_c are not concurrent, then p_{ab}, p_{ac} and L_{bc} are colinear. Both the maps f_{ba} and f_{ca} map L_{bc} to the same point x of L_a. p_{ab}, L_{bc} and x are colinear, as are p_{ac}, L_{bc} and x. By assumption, x is distinct from L_{bc}, so the statement follows.

(2) If no three lines of $L_1, \ldots L_k$ are concurrent, and $p_{12} = p_{13} = p_{23}$ then all the p_{ij} are

equal. It suffices to show that $p_{14} = p_{12}$. By remark (1), $p_{14} \in p_{12}L_{24} \cap p_{13}L_{34}$. These two lines are either distinct, in which case they intersect in $p_{12} = p_{13}$, so $p_{14} = p_{12}$, or the two lines are equal. In this case, p_{12} , L_{24} , p_{13},L_{34} are colinear, and hence on L_4. But then $p_{14} \in L_4$, which is not possible.

(3) If the L_i are as in (2), and $p_{12} = p_{13}$ then all the p_{ij} are equal. $p_{23} \in p_{12}L_{13} \cap p_{13}L_{12}$, and these two lines intersect in $p_{12} = p_{13}$. If the lines are distinct, then $p_{23} = p_{13}$. If they are equal, then the line must be L_1, since it contains two distinct points of L_1. This can not be, for then $p_{12} \in L_1$.

(4) If the L_i are as in (2), and if p_{12} , p_{13} , p_{14} are distinct and colinear, then all p_{ij} are equal. Since $L_{23}\, p_{12}\, p_{13}$ is a line L, and $p_{12} \neq p_{13}$, we have $L_{23} \in L$. Similarly, L_{24} and $L_{34} \in L$. Then $L = L_{23}L_{24} = L_2$ and $L = L_{23}L_{34} = L_3$ - contradiction.

(5) If the L_{ij} are as in (2), and p_{12} , p_{13} , p_{14} are colinear, then all p_{ij} are equal. If the three points are distinct, then (4) applies. If two of them are equal, then (3) and (2) apply.

(6) If L_1 , L_2 , L_3 are concurrent, if there are at least two other lines that do not meet $L_1 \cap L_2 \cap L_3 = x$, if there is no point y such that every line meets x or y, and if $p_{12} = p_{13}$ then all p_{ij} are equal. Suppose we have a mapping α from the union of the L_i to L_1 and on L_2 and L_3 it is given by the perspectivity with center p_{12}. Any line K not passing through $L_1 \cap L_2$ meets $L_1 \cup L_2 \cup L_3$ in three points, and so α is uniquely determined on K. Let K and H be the two lines of the assumption. α is known on K , H , L_1 , L_2 , L_3. Any other line J meets at least three points of these lines, and so α extends uniquely to J. Consequently, α is the perspectivity with center at p_{12}.

(7) A vertex of B(S;PGL(2,q)) is a set of vertices $(x_1, \cdots x_r)$ where for all lines L_i, we have that $x_i \in L_i$ and x_i and x_j are colinear with p_{ij}. Suppose that (x_1,x_2,x_3,x_4) is a vertex and all the x_i are colinear. Then either all p_{ij} are equal or, after renaming,

$x_1 = x_2$ and $x_3 = x_4$. Assume that the x_i are distinct. x_1,x_2 , p_{12} and x_1,x_3 , p_{13} are the same line L, and so $p_{12},p_{13} \in L$. Similarly, $p_{14} \in L$, and so all p_{ij} are equal. If x_1 , x_2 , x_3 are distinct, then p_{23} , p_{24} , p_{34} all lie on L and so by (5) all p_{ij} are equal. If $x_1 = x_2 = x_3$ then since $x_i \in L_i$, L_1 , L_2 , L_3 all intersect at x_1. We conclude that $x_1 = x_2$ and $x_3 = x_4$.

(8) If the L_i are as in (2), and all the points x_i of a vertex of B(S;PGL(2,q)) lie on a line, then all the p_{ij} are equal, for k > 4. Consider any four lines. Either their p_{ij} are equal, or any set of four x's split into two sets of equal pairs. This implies that all the x_i are equal, but then all the lines are concurrent.

(9) Let the L_i be any set of lines. If $p_{ij} \in L_k$ then L_i , L_j , L_k are concurrent. p_{ik} , p_{ij} , L_{jk} is a line L. Since $p_{ij} \in L_k$ and $p_{ij} \neq L_{jk}$ (else $p_{ij} \in L_j$), L is L_k. But then $p_{ik} \in L_k$ so by (1) the L's are concurrent.

(10) If the lines L_i are as in (2), then no p_{ij} lies on any line L_k. This follows immediately from (9).

(11) For any set of L_i, the p_{ij} 's and the L_{ij} 's are distinct. Suppose $p_{ij} = L_{12}$. By (9), L_i , L_j , L_1 and L_i , L_j , L_2 are concurrent. Consequently, $L_{12} = L_{ij}$. But then $p_{ij} \in L_i$ - a contradiction.

We are now in a position to prove some theorems.

Theorem 3

If S is a set of at least 5 lines, no three concurrent, then

$$B(S;PGL(2,q)) \cong S^c \coprod \text{ other colorings}$$

Proof:

This follows immediately from (8) above. □

We conjecture that if there are at least six lines, no three concurrent, then there are no other colorings.

Theorem 4

If every point of the plane meets two lines of S , then B(S;PGL(2,q)) = Ø.

Proof:

By assumption, every point of the plane is an L_{ij}. By (11) above, there are no possible choices for for the p_{ij}. □

If we let L consist of all the lines of the plane missing a point p, then every point except one (p) of the plane is in at least two lines, and there is a coloring.

Theorem 5

If L^c consists of k points on a line ,k < q, then B(L;PGL(2,q)) $\cong L^c$

Proof:

Every point of L lies on two lines, since k < q, so all the p_{ij} lie on the line K which contains the k points. Take three lines L_1 , L_2 , L_3 passing through one of the k points. By (6) if $p_{12} = p_{13}$ then all p_{ij} are equal. There are q different p_{1j}'s, and only k possible values for them, so some two must be equal. Consequently, all the p_{ij} are equal. □

This is false over S_n. If q is a square, we have seen that there are non-standard colorings if k is $\sqrt{q}+1$.

Theorem 6

If L consists of k lines through a point, k< q , then L is PGL(2,q) reflexive.

Proof:

Let the k lines be $L_1, \cdots L_k$ and let α be a coloring of B(L,PGL(2,q)). $B(L,PGL(2,q))$ contains L^c but is much larger. We apply Theorem 5 to find that $B(L^c,PGL(2,q)) \cong L$, so we may assume that $\alpha|\,\mathrm{Im}(\beta)$ is given by the intersection with a line of L, which we take to be L_1. After applying a perspectivity, we assume that

$\alpha(f) = f \cap L_1$, if f is a vertex of $B(L,PGL(2,q))$ which is in the image of β.

Let the vertex f be $(x_1, \cdots, x_k)$. A coloring of L is determined by any points $p_{12} \cdots p_{1k}$ such that $p_{1j} \notin L_1 \cup L_j$. Choose a point p on L_1 distinct from x_1 and $L_1 \cap L_2$. There are q+1 lines through p and k x_i's, so there is some line H through p missing all the x_i's. Let $p_{1j} = H \cap x_1 x_j$. Clearly $p_{1j} \notin L_1 \cup L_j$, so these choices determine a coloring γ of L.

$p_{1j} x_1 x_j$ is a line, so f is a vertex of γ. The vertex g of γ containing $H \cap x_1$ lies in the line H, and so is in Im(β). Consequently, $\alpha(g) = H \cap L_1 = p$ and so $\alpha(f) \neq p$.

If we repeat the construction for the q-1 possible p we see that the only choice for $\alpha(f)$ is $f \cap L_1$ or $L_1 \cap L_2$. Since the latter point is joined to all vertices of B(L,PGL(2,q)), we have $\alpha(f) = f \cap L_1$. □

In the S_n case, this is a trivial result. As a simplicial complex, L is k copies of Δ^q joined at a point v so $B^2(L) \cong B^2(k\Delta^{q-1} * v) \cong B^2(k\Delta^{q-1}) * v \cong k\Delta^{q-1} * v \cong L$.

Theorem 7

If S consists of 3 non-concurrent lines then S is PGL(2,q)-reflexive.

Proof:

β is 1-1, so it suffices to show that every coloring of Im (β) extends uniquely to B(S;PGL(2,q)). A coloring α in Im(β) is determined by some point not in S. A vertex of α is the intersections of a line through p with S. Either we have three colinear points ,or two points, one of them the intersection of two of the lines of S.

Consider a coloring γ not in Im(β), determined by three points p_{12} , p_{13} , p_{23}, distinct and non-colinear. The vertex of γ containing L_{ij} contains just two members. Such a vertex is also determined by a coloring in Im(β). Consequently, every top simplex of B(S;PGL(2,q))-Im(β) meets Im(β) in at least three points. Since a projectivity is determined by three points, the result follows. □

Again, this is an easy result over S_n. We conclude this section with an example. Let S be four lines, no three colinear, in PG(2,3). Since S is the dual of a conic, and all conics are equivalent, any choice of the lines gives the same result. In Figure 1a, we choose A = (1,0,0) , B = (1,0,1) , C = (1,0,2) , D = (0,1,2), E = (0,1,0) , F = (1,1,0) , G = (1,2,0) , H = (1,2,2), I = (1,0,2). Figure 1b gives the simplicial structure of S. B(S;PGL(2,3)) = B(S) and is shown in Figure 1c. There are three points not on L : (1,2,1) , (1,1,1) , and (1,1,2). A calculation gives four colorings of S : all p_{ij} equal to (1,1,1) ; to ((1,2,1) ; to (1,1,2) ; and $p_{12} = (1,2,1) = p_{14}$, $p_{13} = (1,1,1) = p_{24}$, and $p_{23} = (1,1,2) = p_{14}$. The map $\beta : S^c \to B(S)$ is neither onto vertices nor tetrahedra. The central tetrahedron in Figure 1c is the non-standard coloring, which has three vertices in $\mathrm{Im}(\beta)$.

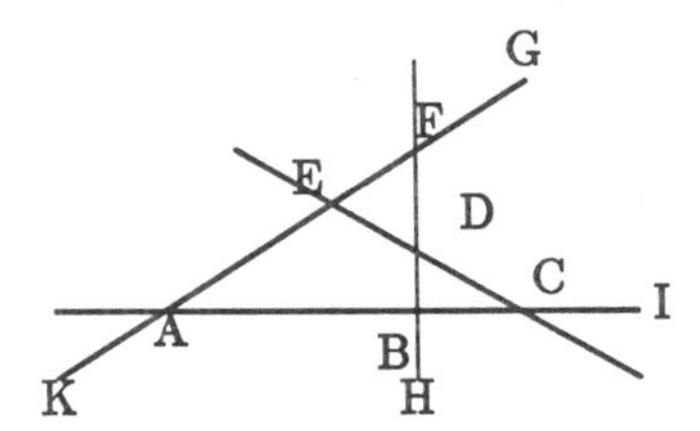

Figure 1a

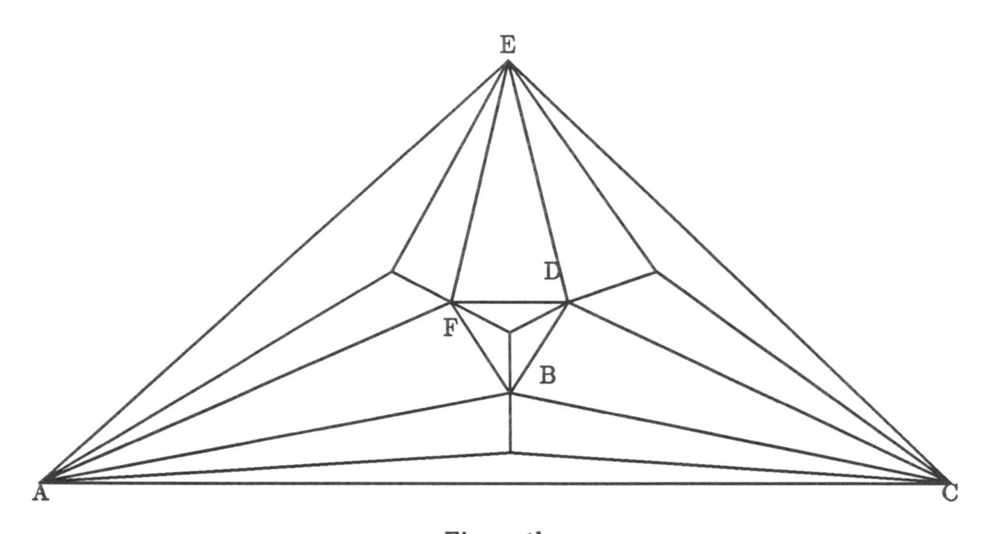

Figure 1b

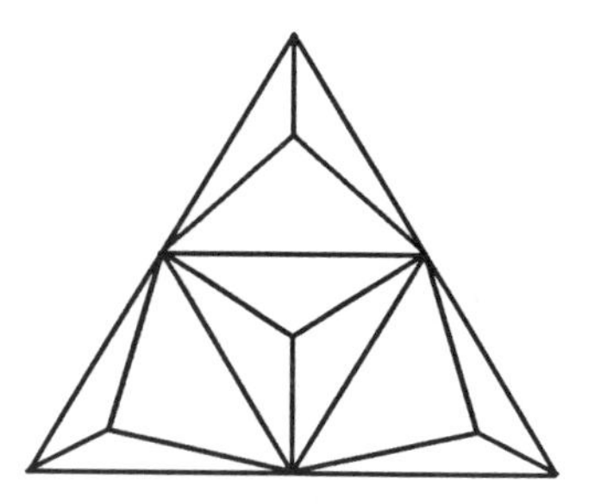

Figure 1c

5. The Hyperbolic Quadric

In Chapter 3.3, we studied the complex X whose top simplices were the lines of a hyperbolic quadric, and whose vertices were the points on these lines. X has a PGL(2,q) structure, as does X^c. Furthermore, $X \cong \Delta^q \# \Delta^q$. Consequently, B(X;PGL(2,q)) consists of all the PGL(2,q)-Latin squares. B(X;PGL(2,q)) has $|\mathrm{PGL}(2,q)| = q(q^2-1)$ vertices. In Chapter 3.3 we calculated that X^c has q^3-q vertices, and so the map $\beta : X^c \to$ B(X;PGL(2,q)) is bijective on vertices.

Is β an isomorphism? We found that X^c had $q^2(q-1)^2/2$ top simplices. Unfortunately, B(X;PGL(2,q)) is sometimes larger.

For q = 2, β is an isomorphism. For q = 3, PGL(2,3) $\cong S_4$, so B(X;PGL(2,3)) $\cong \mathrm{B}(\Delta^3 \# \Delta^3; S_4) \cong S_4$. This has 24 top simplices, whereas X^c has only 18. If q is 4, then PGL(2,4) $\cong$ Alt(5) and this has 72 top simplices, as does X^c. Perhaps β is an isomorphism for even q.

6. Elliptic Involutions of PGL(2,q)

In the study of S_n colorings, there were two recurring families of complexes : $L(K_{2n})$ and $L(K_{n,n})$. For G-colorings, we still have the latter, since it is isomorphic to $\Delta^{n-1} \# \Delta^{n-1}$. In this section we investigate an analog of the former for PGL(2,q).

Recall that $BL(K_{2n})$ can be identified with the permutations of order two with no fixed points. $L(K_{2n})$ is then recovered by applying B. By analogy, consider the set I(q) of elements of PGL(2,q) which have order two and no fixed points. Such maps are called elliptic involutions. For q even, there are none; for q odd there are q(q-1)/2 of them [Hirschfeld, 1979, p. 127]. Give I(q) the graph structure induced as a subgraph of S_n. Equivalently, x and y in I(q) are adjacent iff $xy^{-1} = xy$ has no eigenvectors. We will show that I(q) does have colorings, but the dimension of I(q) appears to be smaller than q-1. The dimension can not be q, for the identity map is disjoint from all elements of I(q). In fact, we show that for q = 3,5,7 the dimension is (q+3)/2. It follows that B(I(q)) is not as interesting as we hoped.

From [Hirschfeld,1979] all elements of I(q) are conjugate to <0,v> = $\begin{pmatrix} 0 & 1 \\ v & 0 \end{pmatrix}$ where v is a non-square in GF(q), q odd. Consider the upper triangular subgroup UT(q) = $\{\begin{pmatrix} 1 & a \\ 0 & 1 \end{pmatrix} \mid a \in GF(q)\}$. Let the conjugate of <0,v> by this element of UT(q) be <a,v>.

We see that <a,v> = $\begin{pmatrix} av & 1-a^2v \\ v & -av \end{pmatrix}$. The <a,v> are all distinct, by an easy calculation. Since there are q(q-1)/2 of the <a,v>, every element of I(q) has a unique representation as <a,v>.

The criterion for <a,v> and <b,w> to be adjacent is that $\langle a,v\rangle\langle b,w\rangle$ has no eigenvalues. The characteristic equation of this product is $\lambda^2+(v+w-vw(a-b)^2)\lambda+vw$ and hence the discriminant is $(v+w-vw(a-b)^2)^2-4vw$. Note that <a,v> is disjoint from <b,w> iff for all c we have that <a+c,v> is disjoint from <b+c,w>.

In case a=b the discriminant is $(v-w)^2$. Since this is a square, there is an eigenvalue, and so the set $\{\langle a,v\rangle \mid v \text{ a non-square}\}$ is an independent set of vertices of I(q). Consequently, the mapping $I(q) \to Z_q$ given by $\langle a,v\rangle \to a$ is a coloring of the graph I(q) with q colors. However, the chromatic number of I(q) is not known to be q.

For <a,v> to be disjoint from <b,v>,the discriminant must be a non-square. If we let t = a-b, then the discriminant is $(2v-v^2t^2)^2-4v^2$. This is a non-square iff $-(4v-v^2t^2)$

is a non-square.

Lemma 1

There are exactly (q-1)/2 t such that $-(4v-v^2t^2)$ is a square. 0 is a value for t iff $t \equiv 3 \bmod 4$.

Proof:

This is a simple algebraic calculation and is omitted.

Lemma 2

I(q) is regular of degree $(q^2-1)/4$.

Proof:

The set of all t in I(q) such that $t(0) = \infty$ is easily seen to be the set of all <0,v>, for v a non-square. Any two members of I(q) meet in at most one transposition. If they met in more, then they would have the same images at four points, and would be identical by the Fundamental Theorem of Projective Geometry. Consequently, given any member of I(q) and any one of its (q+1)/2 transpositions, there are (q-3)/2 members of I(q) which share that transposition. Therefore, each member of I(q) is disjoint from $q(q-1)/2 - ((q+1)/2)\cdot(q-3)/2\ ) -1 = (q^2-1)/4$. □

We conjecture that a maximal set of mutually adjacent vertices of I(q) has at most (q+3)/2 elements. We now see that this is true for q = 3,5,7 by computing I(q).

q=3 I(3) has 3 elements <0,2>, <1,2> and <2,2>. By Lemma 1, the only value of t such that <a,2> and <a+t,2> are not adjacent is t=0. Consequently, $I(3) \cong K_3$. We calculate that $\chi(I(3)) = 3$ and $B(I(3)) \cong \Delta^2$.

q=5 There are 10 elements <i,2> and <i,3>, for i = 0,1,2,3,4. From the Lemma, t= 1,-1 works for the <i,2> and t = 2,-2 works for the <i,3> . The discriminant when a-b=t, u=2 and v = 3 is t^4+1. Consequently, <i,2> is adjacent to $\langle j,3\rangle$ for all $i \neq j$. Knowing all the edges, it is easy to see that I(5) is the complement of the Petersen graph. I(5) contains many sets of 4 mutually adjacent points, but no sets of 5. We have $\chi(I(5)) = 5$ and $B_5(I(5)) \cong \Delta^4$.

q=7 Similar calculations show that the set of all edges of I(7) is

$$\begin{array}{l}
\langle i,3\rangle -- \langle i\pm 2,3\rangle \\
\langle i,3\rangle -- \langle i\pm 3,3\rangle \\
\langle i,5\rangle -- \langle i\pm 1,5\rangle \\
\langle i,5\rangle -- \langle i\pm 2,5\rangle \\
\langle i,6\rangle -- \langle i\pm 1,6\rangle \\
\langle i,6\rangle -- \langle i\pm 3,6\rangle \\
\langle i,3\rangle -- \langle i\pm 1,5\rangle \\
\langle i,3\rangle -- \langle i\pm 2,5\rangle \\
\langle i,3\rangle -- \langle i\pm 2,6\rangle \\
\langle i,3\rangle -- \langle i\pm 3,6\rangle \\
\langle i,5\rangle -- \langle i\pm 1,6\rangle \\
\langle i,5\rangle -- \langle i\pm 3,6\rangle
\end{array}$$

The largest set of mutually adjacent points is seen to be 5. A useful fact for checking these results is that $\langle 0,u\rangle$ is disjoint from $\langle t,v\rangle$ iff $\langle 0,a^2u\rangle$ is disjoint from $\langle a^{-1}t,a^2v\rangle$. delim

7. The Extension Problem

One of the basic results about S_n is that any collection of automorphisms which are disjoint are contained in an (n-1) simplex of S_n. Equivalently, this is the classic result that every partial Latin square extends to a Latin square. In such generality, the analog with an arbitrary structure group is probably false. In this section we give examples for Alt(6) and PGL(2,5) of Latin squares with all rows but one in G. This shows that we can not extend n-1 rows of G to n rows of G.

For Alt(6), here is an even stronger example of a Latin square which has all columns even permutations, and all but the first row an even permutation.

1	2	3	4	6	5
2	1	5	6	4	3
3	4	1	2	5	6
4	5	6	1	3	2
5	6	4	3	2	1
6	3	2	5	1	4

We next give an example for PGL(2,5) . The top row lists the elements of PG(1,5). The left column contains the matrices of PGL(2,5) which when applied to the top row, give the row the matrix is in. The last row is easily checked to not be given by any matrix.

	(1,0)	(1,1)	(1,2)	(1,3)	(1,4)	(0,1)
$\begin{pmatrix} 1 & 1 \\ 1 & 2 \end{pmatrix}$	(1,1)	(1,4)	(1,0)	(1,3)	(0,1)	(1,2)
$\begin{pmatrix} 1 & 2 \\ 2 & 1 \end{pmatrix}$	(1,2)	(1,1)	(0,1)	(1,0)	(1,4)	(1,3)
$\begin{pmatrix} 1 & 3 \\ 3 & 2 \end{pmatrix}$	(1,3)	(1,0)	(1,1)	(0,1)	(1,2)	(1,4)
$\begin{pmatrix} 1 & 4 \\ 4 & 4 \end{pmatrix}$	(1,4)	(0,1)	(1,3)	(1,2)	(1,0)	(1,1)
$\begin{pmatrix} 1 & 0 \\ 0 & 2 \end{pmatrix}$	(1,0)	(1,2)	(1,4)	(1,1)	(1,3)	(0,1)
*	(0,1)	(1,3)	(1,2)	(1,4)	(1,1)	(1,0)

Remember that the reason why we want extension is that if we have extension and the Helly property, then we have reflexivity. In place of the general extension problem, we would be satisfied with the following simplest possible version:

For G = Alt(n), PGL(n,q), or PSL(n,q), is it true that any two disjoint permutations of G are contained in a G-Latin square?

In the remainder of this section we give an example of a group G where extension fails, yet G is reflexive. Let G be the group of semi-linear transformations of the affine line over $GF(p^2)$. The pairs of transformations $x \rightarrow ax+b$, $x \rightarrow cx+d$ and $x \rightarrow ax^p+b$, $x \rightarrow cx^p+d$ are disjoint iff a=c. In each case the pairs belong to a top simplex. However, if the two transformations are $x \rightarrow ax+b$ and $x \rightarrow cx^p+d$ it is possible for them to be disjoint. Since all top simplices are of the form $ax^\sigma+by^\tau+c$ for σ and τ automorphisms of $GF(p^2)$ [Fisk,1983], it follows that such a disjoint pair is not part of a top simplex over G.

It is interesting to note that there are top simplices in S_{p^2} all of whose vertices are semi-linear transformations, but the simplices are not in G. We can construct all such top simplices. Choose a and c such that $V = \{ax-cx^p \mid x \in GF(p^2)\}$ is a 1-dimensional vector space. It is not hard to show that there are $(p^2-1)(p+1)$ such a and c. Choose β_i and γ_j such that they are distinct in $GF(p^2)/V$. Every top simplex has the form

$$x \rightarrow ax+v_k+\beta_i \text{ or } x \rightarrow cx^p+v_k+\gamma_j$$

where the v_k are all the elements in V . There are exactly $(p^2-1)(p+1)(2^p-2)+2(p^2-1)$ top simplices in S_{p^2}.

8. AF(n,q) and bi-affine maps

The affine group AF(n,q) consists of all affine maps f(x) = Ax + b where x and b are in an n-dimensional vector space over GF(q), and A is a non-singular n by n matrix over GF(q). A simplex of Aut(AF(n,q)) is a map

$$F(x,y) = A(x)y + B(x) = \alpha(y)x + \beta(y)$$

where $A(x)$ and $\alpha(y)$ are invertible n by n matrices, and $x,y,B(x),\beta(y)$ are vectors. Such a map F we call **bi-affine** and the study of AF(n,q) requires the knowledge of all bi-affine maps.

We can make some simple reductions in the basic equation. If we let x=0 we get $\beta(y) = A(0)y+B(0)$; letting y=0 gives $B(x) = \alpha(0)x+\beta(0)$. $B(0) = \beta(0)$, so we may assume

$$A(x)y + \alpha(0)x = \alpha(y)x + A(0)y$$

If we let $\tilde{x} = \alpha(0)x$; $\tilde{y} = A(0)y$; $\tilde{A}(\tilde{x}) = A(x)A(0)^{-1}$; and $\tilde{\alpha}(\tilde{y}) = \alpha(y)\alpha(0)^{-1}$ then we have

$$\tilde{A}(\tilde{x})\tilde{y} + \tilde{x} = \tilde{\alpha}(\tilde{y})\tilde{x} + \tilde{y}$$

There is an interesting example, due to George Bergman, which is as follows. Let n = 3 and set $F(X,Y) = X \times Y + X + Y$ where the product is the cross product. If $X = (x_1,x_2,x_3)$, $Y = (y_1,y_2,y_3)$ then

$$F(X,Y) = \begin{pmatrix} 1 & y_3 & -y_2 \\ -y_3 & 1 & y_1 \\ y_2 & -y_1 & 1 \end{pmatrix} \begin{pmatrix} x_1 \\ x_2 \\ x_3 \end{pmatrix} + Y = \begin{pmatrix} 1 & x_3 & -x_2 \\ -x_3 & 1 & x_1 \\ x_2 & -x_1 & 1 \end{pmatrix} \begin{pmatrix} y_1 \\ y_2 \\ y_3 \end{pmatrix} + X$$

The remaining condition is that the matrices are invertible. The determinant is $1+x_1^2+x_2^2+x_3^2$ and a similar formula holds for the y's. For any field of non-zero characteristic, one can find values of x_1,x_2,x_3 which make the determinant zero. On the other hand, this is a good example for any ordered field, in particular the reals.

This example actually arises from quaternionic multiplication; one can also get a bi-affine map of non-standard form over the reals by using the Cayley algebra in dimension 8. It is conjectured that over the reals these are all the examples.

If there were no examples over GF(q) other than $F(x,y) = Ax + \alpha y+b$ then we would have that $AF(n,q) \cong |GL(n,q)| \bullet \Delta$. We know of one counterexample. Since $AF(2,2) \cong$ Sym(4) as a group, and AF(2,2) is a permutation group on a set of four elements, $AF(2,2) \cong S_4$ as a complex. It is not the case that $S_4 \cong 6 \bullet \Delta^3$.

In S_4, any two top simplices meet in at most 2 points. This is a consequence of the following result, whose proof is omitted.

Theorem 1

Any two intersecting top simplices of AF(n,q) meet in exactly q^i elements for some i $0 \leqslant i \leqslant n$.

This identification gives us a nice representation of S_4. The following figure shows the verticies of $K_{3,3}$ labeled with the six members of GL(2,2). Each edge is labeled with a member of the two-dimensional vector space over $\mathbb{Z}_2$. Each vertex labeled α determines a tetrahedron of S_4: $\{\alpha , \alpha+(1,0) , \alpha+(0,1) , \alpha+(1,1)\}$. An edge labeled a joining verticies labeled α and β determines two tetrahedra. If b and c are the other two non-zero vectors of $\mathbb{Z}_2 \oplus \mathbb{Z}_2$, then the tetrahedra are $\{\alpha , \alpha+a, , \beta+b , \beta+c\}$ and $\{\beta , \beta+a, , \alpha+b , \alpha+c\}$. To see how this works, suppose $\alpha(x) = \beta(x)+b$. This happens iff b is in the image of $\alpha + \beta$. However, the labeling is chosen so that the image of $\alpha + \beta$ contains only one non-zero element: a.

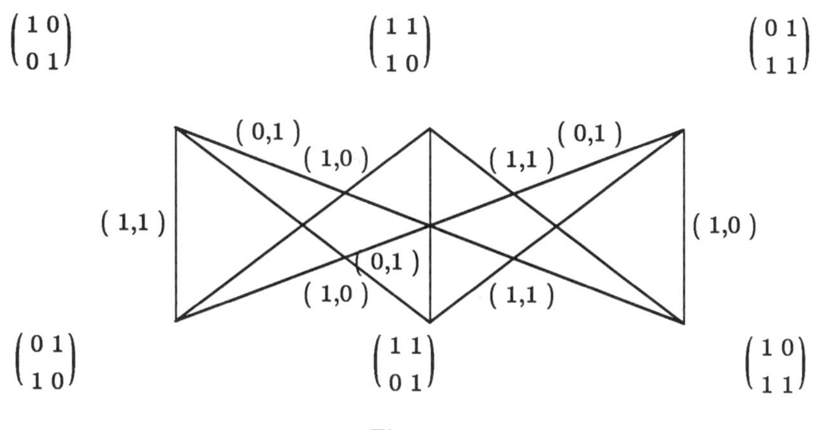

Figure 1

Reflexive and Self-Dual Complexes

This Chapter investigates several classes of reflexive and self-dual complexes. In the first section we list the known self-dual complexes. We introduce a recursive class of complexes called binary n-trees and show that they are all reflexive. In the following section we show that the only reflexive bipartite graphs are disjoint unions of P_{2n}'s and paths. We then show that all the members of a certain recursive class of triangulations of the disk are all reflexive. We next give a construction which associates a reflexive 2-complex to every reflexive triangulation of the disk. We end the Chapter by showing that there is an infinite class of triangulations of the disk which are not reflexive.

1. The Color Spectrum

The **color spectrum** is the set of all integers k for which there is a connected complex X satisfying B(X) = kX. If k is 1, the corresponding X are the self-dual complexes. So far, we only know that 1 and 2 are in the color spectrum.

If X and Y are self-dual complexes, then B(X∗Y) = B(X)∗B(Y) = X∗Y, so X∗Y is also self-dual. All the known self-dual complexes are Δ^0, $2\bullet\Delta^1$, $L(K_6)$, the complexes in Figure 1, and the joins of any number of them. Since Δ^n is the join of n+1 copies of Δ^0, Δ^n is self-dual.

There is an infinite family of complexes with k=2, and one other one. We saw in Chapter 2, that $B(S_n\#\Delta) \cong 2\bullet S_n\#\Delta$. The other is the line graph of a certain graph C known as the Coxeter graph [Coxeter, 1980]. One construction for C is as follows. The vertices are all unordered pairs of distinct vertices of PG(1,7). Two vertices (a,b) and (c,d) form an edge iff (a,b;c,d) is a harmonic set. C is a very special graph with many other unusual properties.

If we drop the assumption of connectivity in the definition of the color spectrum, then the color spectrum is infinite. Suppose that B(X) = kX, and let Y = $n\bullet X^{\times n}$. B(Y) = $(\mathrm{B}(X^{\times n}))^{\times n} = (n\bullet B(X))^{\times n} = (nk\bullet X)^{\times n} = n^n k^n X^{\times n} = n^{n-1}k^n\bullet Y$. Since 1 and 2 are in the color spectrum, n^{n-1} and $n^{n-1}2^n$ are in this modified color spectrum.

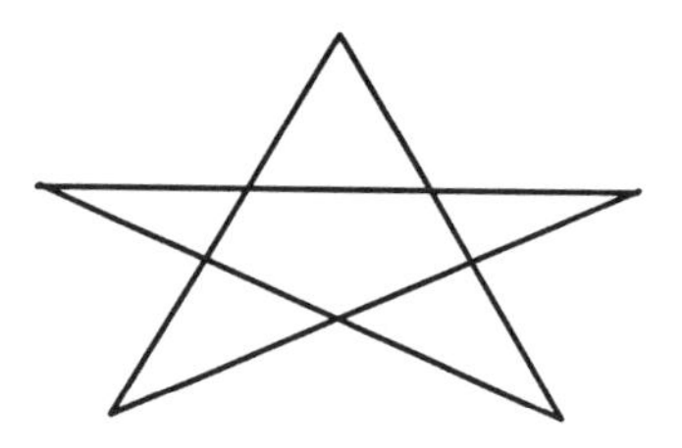

Figure 1a

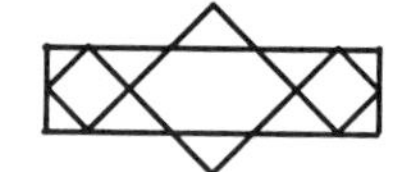

Figure 1b,c

Figure 1d,e (hatting omitted)

2. Binary n-Trees

In this section we introduce the class of binary n-trees, and show that they are all reflexive. We define a binary n-tree recursively. An n-simplex is a binary n-tree. If T is a binary n-tree, and σ is an (n-2)-simplex of T which is contained in exactly one n-simplex of T, we can form a new binary n-tree T$'$ by adding an n-simplex which meets T in exactly σ. Figure 1 gives an example of a binary 2-tree.

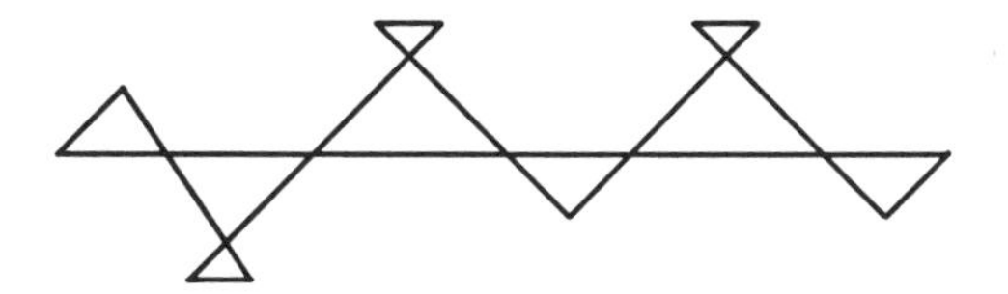

Figure 1

If σ is a simplex of a complex T, define $\rho(\sigma, T)$ to be the number of n-simplices containing σ.

Theorem 1

Let X be an n-complex, and σ an (n-2)-simplex of X. Form a new complex Y by adding to X an n-simplex T which meets X exactly in σ. Then

(a) If $\rho(\Phi(\sigma), B^2(X)) = 1$ then $B^2(Y)$ is formed from $B^2(X)$ by adding an n-simplex to $\Phi(\sigma)$.

(b) If $\rho(\Phi(\sigma), B^2(X)) = 1$ and X is reflexive, then Y is reflexive.

Proof:

Let the two vertices of T-σ be labeled 1 and 2. Let σ lie in the n-simplex D of X, and let the vertices of D-σ be labeled 4 and 5. Every coloring of Y corresponds to a coloring of G along with a coloring of $D \cup T$. We can thus write B(Y) = $A_1 \cup A_2$, where A_1 (resp. A_2) is all the colorings f of Y with f(1) = f(5) (resp. f(1) = f(4)). $A_1 \cong A_2 \cong B(G)$ and $A_1 \cap A_2 = \Phi(\sigma)$.

Since $\rho(\Phi(\sigma), B^2(X)) = 1$ the complement of $\Phi(\sigma)$ in B(X) consists of 1 connected bipartite graph containing all the vertices of B(X)-$\Phi(\sigma)$. If f is coloring of B(G), then knowing the values of f on $\Phi(\sigma)$ determines f on all of B(G), unless f is a 2-coloring on B(X)-$\Phi(\sigma)$. In the latter case there are two possible f's. Consequently, all colorings of

B(Y) correspond uniquely to a coloring of B(X), unless the coloring of f uses exactly n-1 colors on $\Phi(\sigma)$. In this case there are two different colorings of B(Y) and these correspond to $\Phi(T)$ and $\Phi(D)$. □

Corollary 2

(a) A binary n-tree is reflexive

(b) If we attach (as in Theorem 1) a binary n-tree to a reflexive complex, the resulting complex is reflexive.

If the simplex σ does not have $\rho(\Phi(\sigma),B^2(X)) = 1$ then Y may not be reflexive even if X is. For example, $\hat{P_4}$ is reflexive, but if we add a triangle to one of the points of degree two (Figure 2a), we do not get a reflexive complex.

Figure 2a,b

Corollary 3

If X and Y are as in Theorem 1, and if $\rho(\Phi(\sigma,B^2(Y)) = 2$ and Y is reflexive, then $\rho(\Phi(\sigma,B^2(X)) = 1$ and X is reflexive.

Proof:

In general, if τ is a (n-2)-simplex of B(Z), for some n-simplex Z, then $\rho(\Phi(\tau,B(X)) = 2^{n-1}$ where X-τ has n components. X-τ means the graph determined by all vertices not in τ. If X and Y are as in Theorem 1, then if $\rho(\Phi(\sigma,B^2(X)) = 2^{n-1}$ then $\rho(\Phi(\sigma,B^2(Y)) = 2^{2n-1}$. Since B(X)-$\Phi(\sigma)$ consists of n connected components, B(Y)-$\Phi(\sigma)$ has 2n connected components. This relationship of degrees implies that the degree of σ in $B^2(X)$ is 1. We now use Theorem 1 to conclude X is reflexive. □

Corollary 4

If Y is X plus a binary n-tree (joined as in Theorem 1) then X is reflexive iff Y is reflexive.

3. Reflexive Bipartite Graphs

If G is a graph, we say that G is reflexive if $B^2(\hat{G}) \cong \hat{G}$. The fact that we only use three colors allows us to prove a characterization theorem for reflexive bipartite graphs.

Theorem 1

If G is bipartite, then for all $p \in V(G)$, $\rho(\Phi(p),B^2(\hat{G})) \geqslant 2^{\rho(p)-1}$.

Proof:

The degree of $\Phi(p)$ in $B^2(\hat{G})$ is 2 to one less than the number of connected components of $B(G)-\Phi(p)$. We first show that in any connected component H of $B(G)-\Phi(p)$, the vertices adjacent to p are always colored the same. Let f and g be adjacent colorings in $B(G)-\Phi(p)$. Let the colors be 1, 2, 3. If f were obtained from g by switching a 2-3 cycle, then the set v of points colored 1 by f and g is the common vertex of f and g in B($\hat{G}$). Since $v \in \Phi(p)$, we would have that f and g are not adjacent in $B(G)-\Phi(p)$.

We may now assume that g is obtained from f by switching some 1,3 cycle. But no vertex colored 3 that is adjacent to p can be in the cycle, for then p would be changed as well. No vertex colored 2 changes, so all the points adjacent to p are the same color in H.

The above is generally true, and no information about G was used. If G is bipartite, there is a coloring of G using just two colors 1 and 2. By replacing the color of any subset of vertices adjacent to p by the color 3, we get $2^{\rho(p)-1}$ different colorings of the vertices adjacent to p, and hence at least that many components. $\square$

Theorem 2

If G is a finite bipartite graph satisfying $B^2(\hat{G}) \cong \hat{G}$, then G is a disjoint union of intervals and circles on an even number of points.

Proof:

If some vertex p in V(G) has degree greater than 2, then $\rho(\Phi(p), B^2(\hat{G})) \geqslant 2^{\rho(p)-1} > \rho(p)$, so G is not reflexive. The only such bipartite graphs are the ones stated in the conclusion. $\square$

There are other graphs which are reflexive, namely circles on an odd number of vertices. We do not know if there are any others.

4. Sparse Planar Triangulations

In this section we introduce a class of complexes called sparse planar triangulations and show that they are all reflexive. The method proof is inductive, using direct limits. We give a recursive definition of sparse planar triangulations.

(1) Δ^2 is a sparse planar triangulation.

(2) If G is a sparse planar triangulation and p is a point of the boundary of G which lies in exactly two triangles Abp and acp, then the result of adding a new vertex q and triangles qbp and qcp is a sparse planar triangulation.

(3) If G is a sparse planar triangulation and ab is a boundary edge of G contained in triangle abc, where c is an interior point of G, then the result of removing triangle abc and adding a new point q and two new triangles acq and bcq is a sparse planar triangulation.

(4) If ab is a boundary edge lying in the triangle abq of the sparse planar triangulation G then adding a new triangle pab gives a sparse planar triangulation.

(5) All sparse planar triangulations are obtained from (1) by a sequence of constructions of type (2),(3) and (4).

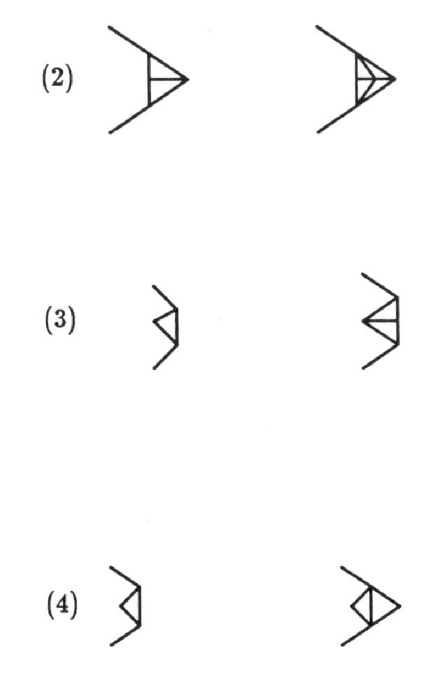

Figure 1

For example, Figure 2 gives a sequence of sparse planar triangulations obtained by applying rules 1,4,2,3,3,4,4,3,3.

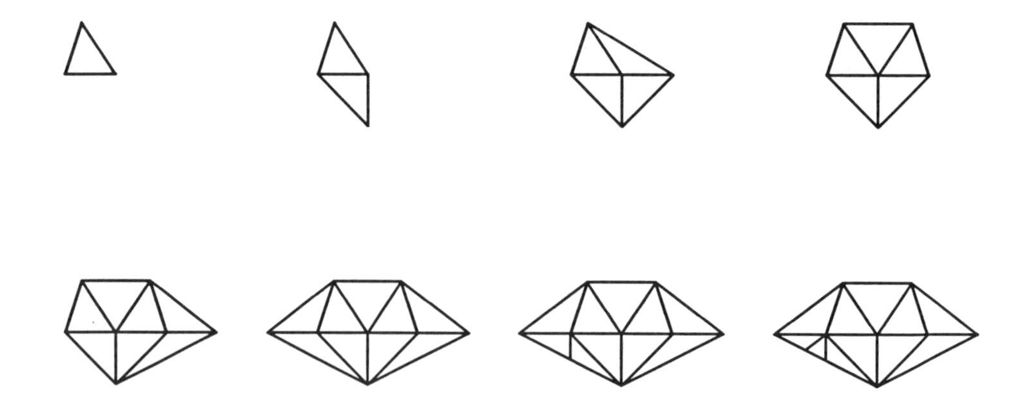

Figure 2

We next show that there is a map from $\hat{G}$ to $\hat{G}'$ for all $\hat{G}$ and $\hat{G}'$ in the definition. In case (2), we map G to G' by sending q to a and all other points to themselves. In case (4), we map p to q and send all other points to themselves. In (2) and (4), we then extend the map to the hatted complex. In case (3), let the points of $\hat{G}'$ over abc be D, over acq be E and over bcq be F. Map E to b, F to a, q to D, and the remaining points to themselves.

In case (2), we say that bp and pc are *fold edges* . In case (3), cq is the fold edge, and in case (4), ab is the fold edge. We call c in case (3) the center. We call the maps the elementary fold maps. If we obtain G from some H by a sequence of constructions of type (2),(3) or (4), then we get a map from G to H given by the composition of the elementary fold maps. The only construction to introduce an interior point in case (2), so all interior points have degree at least 4.

Let X be an arbitrary sparse planar triangulation and let $\{X_i\}$ be the collection of all the sparse planar triangulations such that X can be obtained from X_i by a sequence of operations (2),(3) and (4). This is a directed system with the maps being the fold maps.

Lemma 1

If X is a sparse planar triangulation with at least two triangles, then $\lim\limits_{\rightarrow}(\hat{X}_i,f) = \hat{X}$.

Proof:

Define the fold region to be the set of triangles containing a fold edge. If X has two foldings X_1 and X_2 and the triangles of their fold regions are disjoint, the we can preform both the foldings to obtain X_3. In this case we have $\lim\limits_{\rightarrow}\{\hat{X}_1,\hat{X}_2,\hat{X}_3\} = \hat{X}$.

If X is either $P_4 * p$ or $P_4 * p$ minus one triangle, the Lemma is seen to hold. By induction, if X is larger than either of these triangulations, then we can find two fold edges with disjoint fold regions. It follows that $\lim\limits_{\rightarrow}(\hat{X}_i,f) = \lim\limits_{\rightarrow}\{\hat{X}_1,\hat{X}_2,\hat{X}_3\} = \hat{X}$. □

We do not need all the foldings to conclude that X is a direct limit, but we will need them to show that X is reflexive.

Theorem 2

All sparse planar triangulations are reflexive

Proof:

Write $\lim\limits_{\rightarrow}(\hat{X}_i,f) = \hat{X}$. By induction, we assume that all the X_i are reflexive. Although it is not true that $\lim\limits_{\leftarrow}(B(X_i),Bf) = B(X)$, we will show that they have the same colorings. It will follow that $B^2(X) = B(\lim\limits_{\leftarrow}(B(X_i),Bf) = \lim\limits_{\rightarrow} B^2(X_i),B^2f) = \lim\limits_{\rightarrow}(X_i,f) = X$. Since each of the fold maps $X \rightarrow X_i$ is onto, the maps $B(X_i) \rightarrow B(X)$ are 1-1. We identify $\lim\limits_{\leftarrow}(B(X_i),Bf)$ with the union of the $B(X_i)$ in B(X).

If X can be obtained from Y by a construction of type (4), then by Corollary 2 of Section 2, the reflexivity of Y implies that of X. We may thus assume that X has no fold edges of type (4).

Let f be a coloring of X. If there is a non-singular fold edge e of X of type (2), then f is induced by a coloring of X folded along e. Thus f is in $\varprojlim(B(X_i),Bf)$. If e and e' are fold edges in case (3), and they are singular edges of f, then again f is in $\varprojlim(B(X_i),Bf)$.

We thus assume that f is singular on every fold edge of type (2) and non-singular on fold edges of type (3). Suppose that p is a center which has at least one fold edge of type (2). Then p has at least adjacent two fold edges of type 2. Let a,b,c,d be consecutive edges in the link of p such that f(a)=f(c)=1, f(d)=f(d)=2, f(p)=3. We construct a new coloring by setting F(b)=4. F is non-singular at pc, and so is in $\varprojlim(B(X_i),Bf)$. F and f have vertices $f^{-1}(3)$ and $f^{-1}(1)$ in common, and so are in $\varprojlim(B(X_i),Bf)$. By setting G(c)=4, we see $f^{-1}(2)$ is in $\varprojlim(B(X_i),Bf)$. Thus at least 3 of the four vertices of f are in $\varprojlim(B(X_i),Bf)$.

If X has no fold edge of type 2, we consider a center p and its four neighbors a,b,c,d, with b in the boundary of X. Assume that f(a)=f(c)=1, f(b)=2, f(d)=3, f(p)=4. Define F(b)=3. This shows that $f^{-1}(1)$ and $f^{-1}(4)$ are in $\varprojlim(B(X_i),Bf)$. If we set G(p)=2 and G(b)=4, we see $f^{-1}(3)$ is in $\varprojlim(B(X_i),Bf)$.

We thus see that every tetrahedron of B(X) meets $\varprojlim(B(X_i),Bf)$ in at least three points. It follows that every coloring of $\varprojlim(B(X_i),Bf)$ has at most one extension to B(X). □

5. Edge Coloring 3-Complexes and Reflexive 2-Complexes

In this section we associate a 2-complex M(Y) with every 3-complex Y. Under certain conditions, M commutes with B^2. Applying M to an SPT gives a reflexive 2-complex. This gives an infinite family of reflexive 2-complexes.

If Y is a 3-complex, define an edge coloring of X to be a coloring of the edges of Y with three labels such that every triangle of Y has edges of all three colors. Δ^3 has a unique edge coloring, and under this coloring any two disjoint edges have the same color. If T is a tetrahedron of Y, each pair of disjoint edges of T must have the same color under any edge coloring of Y.

Define two edges of Y to be equivalent if they are disjoint and lie in some 3-simplex of Y. Extend this relation to an equivalence relation. All the edges in an equivalence class are colored alike under any edge coloring of Y.

Define the vertices of M(Y) to be all equivalence classes of edges of Y. Three equivalence classes form a triangle of M(Y) iff there is some triangle of Y which contains one edge from each equivalence class. There are Y for which M(Y) has no triangles, but this will never happen if Y has an edge coloring. If $g : X \to Y$ is a non-degenerate map, then M(g) : M(X) $\to$ M(Y) is well-defined, and is non-degenerate.

For example, take Y to be $\hat{P}_n * p$. This is a wheel with a point added over each triangle. M(Y) is seen to be $\hat{P}_n$.

The edge colorings of Y are in 1-1 correspondence with 3 colorings of M(Y). We can construct edge colorings of Y by inducing them from colorings of Y. If f : Y $\to \Delta^3$ is a coloring, define the edge coloring F(e) to be the color of f(e), where Δ^3 is given its unique edge coloring.

If vw is an edge of B(Y), then there is a coloring f of Y such that v = $f^{-1}(1)$ and w = $f^{-1}(2)$, where we take the four colors to be 1,2,3,4. If we know the set of points colored 1 or 2, then we know the set of points colored 3 or 4. Consequently, all edges of B(Y) in the same equivalence class determine the same edge coloring of X. This gives us a map $M(B(Y)) \to B(M(Y))$.

There is one situation where this map is known to be an isomorphism. Let X be a triangulation of a simply connected surface. $M(\hat{X})$ is the line complex L(X). The edge colorings of $\hat{X}$ are in 1-1 correspondence with the edge colorings of X which is the same as the colorings of L(X) . Since every edge coloring of X is induced by a four coloring [Fisk,1978], we have $M(B(\hat{X})) \cong B(M(\hat{X}))$.

We can now explain the relationship between some of the self-dual examples of Section 1. If X is a triangulation of a disk, and $\hat{X}$ is self dual then $M(\hat{X}) \cong M(B(\hat{X})) \cong B(M(\hat{X}))$. Consequently, $M(\hat{X}) = L(X)$ is self dual. Thus, the two 2-complexes of Section 1 are the result of applying M to the hatted triangulations of Figure 1d,e. Next, we show that under certain conditions M and B^2 commute.

Theorem 1

Suppose that X is a triangulation of a simply connected surface and $B(\hat{X})$ is connected. Then

$$B^2(M(\hat{X})) \cong M(B^2(\hat{X}))$$

Proof:

Since X is simply connected, $B(M(\hat{X})) = M(B(\hat{X}))$. We will show that every edge coloring of $B(\hat{X})$ is induced by a four coloring. Equivalently, $B(M(B(\hat{X}))) \cong M(B^2(\hat{X}))$. The Theorem then follows from $B^2(M(\hat{X})) \cong B(M(B(\hat{X}))) \cong M(B^2(\hat{X}))$.

Suppose that D is a disk of the form $P_n * p$ contained in X. Since $M(\hat{D}) \cong \hat{P}_n$, we have $B^2(M(\hat{D})) \cong B^2(\hat{P}_n) \cong \hat{P}_n \cong M(B^2(\hat{D}))$. Consequently, every edge coloring of $B(\hat{D})$ is induced by a four coloring of $B(\hat{D})$.

An edge coloring of $M(B(\hat{X})$ can be viewed as a specific choice of maps for each coloring such that if f and g share the vertex v, then there is a color c such that $f^{-1}(c) = g^{-1}(c) = v$. Since D is contained in X, this is a coherent collection of colorings of D, and hence is induced by a four coloring.

If X has no vertex whose link is a circle, then all vertices are on the boundary, and X is a SPT obtained by a sequence of single triangle additions. In this case it is easy to see that both $\hat{X}$ and M($\hat{X}$) are reflexive, so we are done. □

If B($\hat{X}$) is not connected, the theorem may be false. Take X to be the icosahedron. $M(B(\hat{X})) = 10 \cdot \Delta^2$, so $B^2(\hat{M}(X)) \cong \Delta^{\times 10}$, but $M(B^2(\hat{X}))$ has only 24 triangles.

Corollary 2

If X is an SPT, then M($\hat{X}$) is reflexive.

6. Reflexive Triangulations of the 2-Sphere

Suppose that X is a triangulation of some 2-manifold. If $\hat{X}$ is reflexive, we say that X is reflexive. The only known examples of reflexive triangulations of the 2-sphere are the tetrahedron, and the triangulations of the form $P_{2n} * S^0$, where P_{2n} is the polygon with 2n vertices, and S^0 is two disjoint points. We conjecture that these are all the reflexive triangulations. We have the following result about even subdivisions.

Theorem 1

If G is a triangulation of the 2-sphere with at least two odd vertices, then every sufficiently fine even subdivision of G is not reflexive.

Proof:

If G has exactly two odd vertices, then they have the same color under any coloring of G (Tutte's Lemma [Fisk, 1978]). Thus, the two odd vertices have the same image under Φ, and so Φ is not one to one. If G has at least four odd vertices, then every sufficiently fine even subdivision H of G has colorings of even and odd degree [Fisk, 1978]. Since the degree of a coloring is determined by any one of its vertices (Tutte's Lemma again), the colorings of even and odd degree are disjoint. If we write B(H) = C$\cup$D, then $B^2(H) = B(C) \times B(D)$. Since no triangulation contains the product of two tetrahedra, H is not reflexive. □

The icosahedron I has an unusual coloring structure. $B^2(\hat{I}) \neq \hat{I}$ but $B(\hat{I}) = B^3(\hat{I})$. We have some partial results about non-reflexive even triangulations. We can show that there are certain triangulations of the disk such that if an even triangulation T contains one of them, then T is not reflexive. This allows a forbidden subgraph approach to determining all reflexive even triangulations.

Define X to be *full* in Y if X is a subcomplex of Y and no edge of Y has both its endpoints in X. Suppose X is an even triangulation of a disk, and Y is an even triangulation of either a disk or 2-sphere. If X is full in Y, then every vertex p of $B(\hat{X})$ extends to a vertex of $B(\hat{Y})$. Indeed, take the extension to be $p \cup (\hat{X} - \hat{Y})$. Since no two points of $\hat{X}$ meet in $\hat{Y}$ that did not already meet in $\hat{X}$, this is a vertex of $B(\hat{Y})$. X full in Y implies that $B(\hat{Y}) \to B(\hat{X})$ is one to one.

Theorem 2

Let Y be a subtriangulation of X . Assume

(1) X has a three coloring

(2) $B(X) \to B(\hat{Y})$ is onto

(3) $\Phi_Y : \hat{Y} \to B^2(\hat{Y})$ is strictly into

then $\Phi_X : \hat{X} \to B^2(\hat{X})$ is strictly into.

Proof:

We have maps $i : \hat{Y} \to \hat{X}$, $i_* : B^2(\hat{Y}) \to B^2(\hat{X})$, $\Phi_X : \hat{X} \to B^2(\hat{X})$ and $\Phi_Y : \hat{Y} \to B^2(\hat{Y})$ satisfying $\Phi_X \circ i = i_* \circ \Phi_Y$. Suppose that p is a vertex of $B^2(\hat{Y})$ which is not in $\Phi_Y(\hat{Y})$. $i_*(p)$ consists of all vertices of $B^2(\hat{X})$ which contain a vertex of $B(\hat{X})$ which contains a vertex of $B(\hat{Y})$ which appears in p. Assume that Φ_X is onto vertices and let $x \in \hat{X}$ satisfy $\Phi_X(x) = i_*(p)$. $\Phi_X(x)$ consists of all vertices of $B(\hat{X})$ which contain x. By hypothesis (2), every vertex of B($\hat{Y}$) contained in p extends to a vertex of B($\hat{X}$) containing x.

If $x \in \hat{Y}$, then by commutativity, $i_* \circ \Phi_Y(x) = \Phi_X(x) = i_*(p)$. Since i_* is one to one, $p = \Phi_Y(x)$, which we assumed to be false. Since $x \notin \hat{Y}$, there is a vertex P of B($\hat{X}$) which contains p. P consists of all the vertices of p along with $\hat{X}-X-\hat{Y}$. It follows that $x \in \hat{X}-X-\hat{Y}$.

Next, suppose that $B^2(\hat{Y})$ contains two adjacent vertices $p,p' \in B^2(\hat{Y})-\mathrm{Im}(\Phi_Y)$. They correspond to two adjacent vertices x,x' of $\hat{X}$. However, x and x' are not adjacent, for they both belong to $\hat{X}-X-\hat{Y}$, which consists of isolated points. If there is no point of $B^2(\hat{Y})-\mathrm{Im}(\Phi_Y)$ adjacent to $\hat{X}$, then x is contained in a tetrahedron whose other vertices are of the form $\Phi_Y(a)$, $\Phi_Y(b)$, $\Phi_Y(c)$. Since X has a three coloring, a ,b c form a triangle in $\hat{Y}$, and so x is the other vertex of the unique tetrahedron containing a,b,c. This implies x is in $\hat{Y}$ - contradiction. □

Corollary 3

Let D be the set of all disks that

(1) have a three coloring

(2) $D \to B^2(D)$ is not onto (w.r.t vertices)

then no reflexive triangulation with a three coloring contains any member of D as a full subcomplex.

The set D is not empty. Figure 1 gives a member of D, which was verified by computer calculations.

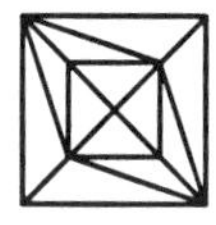

Figure 1

Continuous Colorings

The purpose of this Chapter is to extend the idea of coloring to topological spaces. We begin by showing that Hom,Aut, and B all have the expected properties in this topological category. For the category of pure simplicial complexes, we are able to show that B applied to an infinite product is the disjoint union of B applied to the individual terms. Generalizing earlier work, we show that the infinite analogs of an interval and $\Delta \# \Delta$ are reflexive. Using the finite case as a guide, we define Latin Square spaces, and show that the 2-sphere is not such a space. We end the Chapter with a discussion of colorings of the cartesian product of the reals with the reals. The space is not reflexive, and the exceptional colorings are given by finite compactly supported measures.

1. Continuous Colorings

In this section we introduce topology to our study of colorings. We define continuous colorings and prove various elementary facts about them.

We are going to have colorings with infinitely many colors, so we introduce a different terminology from the rest of this paper. Let Ω be a topological space, and let G be a group of homeomorphisms of Ω. An *G-structure* (T,V) consists of a topological space V and a set T of homeomorphisms from Ω to V. We call the members of T the blocks of (T,V) ; the members of V are called vertices. The homeomorphisms in T are called the structure maps. If t$\in$T, v$\in$V, and v$\in t(\Omega)$, then we say that v is incident with t. If v,w $\in$V, and there is a t in T such that v and w are both incident with t ,then we say v and w are adjacent.

Examples

(1) Suppose that X is a pure simplicial complex of dimension n-1. We associate an S_n-structure (T,V) as follows. Ω is a set of n points with the discrete topology. V is the set of vertices of X, endowed with the discrete topology. T is the set of top simplices. T has one member for each n-1 simplex of X. If t in T corresponds to a top

simplex D of X, then the mapping corresponding to t is any bijection from Ω to D. It is important to note that the topology on V is not the topology of X as a simplicial complex.

(2) Take Ω to be the real projective line PG(1,$\mathbb{R}$). Let G be PGL(2,$\mathbb{R}$), the group of the projective line. Let T be a set of lines in the real projective plane, and let V be the set of points on the lines. We give V the topology induced by the projective plane. We choose a specific line in the plane and identify it with PG(1,$\mathbb{R}$). For each line in T, the corresponding structure map is any projectivity from the line to PG(1,$\mathbb{R}$).

(3) For any set G of homeomorphisms of a space Ω, we have the trivial G-structure which has $V = \Omega$ and T = { identity map }. We denote this by Ω. The two basic questions about a G-structure I that we ask are (1) are there any maps from I to Ω and (2) what is the structure of all maps from I to Ω ?

(4) Let M be a topological n-manifold. Let Ω be the n-disk, and let G be the group of all homeomorphisms of Ω. Let V = M as a topological space, and let T be the set of all homeomorphisms from Ω to M. Since any two points of M are contained in an n-disk, any two points of V are adjacent. This example is not interesting to us, for there are so many structure maps that there are no colorings.

A map between two G-structures (T,V) and (S,W) is a continuous map $f : V \to W$ such that (1) for every t in T there is an s in S such that f is a homeomorphism from $t(\Omega)$ to $s(\Omega)$ and (2) the map $t \circ f \circ s^{-1} : \Omega \to \Omega$ is in G. In other words, the image of every block is a block, and the induced map on Ω is in G. We call f a G-map.

A **coloring** of a G-structure (T,V) is a G-map from (T,V) to Ω. We say that two colorings f and h are equivalent if there is a g in G such that $f = g \circ h$. If we think of the members of Ω as colors, then we say that f and h differ by a permutation of colors.

We are now ready for the most important construction. Let I = (T,V) be a G-structure. The space of colorings of I, B(I;G), is a new G-structure (S,W). S consists of all equivalence classes of colorings of I. The members of W are all subsets of V of the form $f^{-1}(p)$ where $p \in \Omega$ and f is a coloring of I. If f is any coloring of I, we define the map F : $\Omega \to W$ by $F(p) = f^{-1}(p)$. We call $f^{-1}(p)$ a vertex of f. The topology is defined on W by requiring that all structure maps are homeomorphisms onto their images and that for each open set O in V the set $\{f^{-1}(p) \mid f^{-1}(p) \cap O \neq \emptyset\}$ is open in B(I;G).

The second condition comes from requiring that Φ is a G-map. For each t in T there is a map $\Phi(t) : B(I;G) \to \Omega$ defined by $\Phi(t)\{f^{-1}(p)\} = f^{-1}(p) \cap t(\Omega)$. Since f is a coloring, exactly one point of t(Ω) is sent to p, so $\Phi(t)$ is well defined. The requirement for $\Phi(t)$ to be continuous is that for each open set U of Ω, the set $\Phi(t)^{-1}$(U) is open in B(I;G). This implies that $\{f^{-1}(p) \mid f^{-1}(p) \cap t(\mathrm{U}) \neq \emptyset\}$ is open.

We can now define $\Phi : I \to B^2(I;G)$ by sending v∈V to $\{f^{-1}(p) \mid v \in f^{-1}(p)\}$. The last paragraph shows that $\Phi(t)$ is a coloring, so Φ is a G-map.

Examples

(1) Let I = Ω. There is obviously only one equivalence class of maps from Ω to Ω, so B(I;G) = I.

(2) Let $\mathbb{Z}$ denote the simplicial complex whose vertices are the integers . There is an edge from x to x+1 for all integers x. We can define B($\hat{\mathbb{Z}}$) without recourse to this Chapter, but it will turn out that if we include all colorings of B($\hat{\mathbb{Z}}$), then $\hat{\mathbb{Z}}$ is not reflexive. If however, we include only the continuous colorings of B($\hat{\mathbb{Z}}$), then $\hat{\mathbb{Z}}$ is reflexive.

A 3-coloring of $\hat{\mathbb{Z}}$ is uniquely determined by its values on $\mathbb{Z}$. Consequently, a vertex of B($\hat{\mathbb{Z}}$) is equivalent to an independent set of $\mathbb{Z}$. Three independent sets of $\mathbb{Z}$ form a coloring if they are disjoint, and their union is $\mathbb{Z}$. The open sets of B($\hat{\mathbb{Z}}$) have as a basis the sets { all independent sets containing x }, where x $\in \mathbb{Z}$ and {all independent sets not containing x or x+1 }.

(3) Suppose that X is a simplicial complex. A basic open set of B(X) is of the form $U_C = \{f^{-1}(p) | f^{-1}(p) \cap C \neq \emptyset\}$ where C is a set of points in some simplex of X. Since a block of X has a finite number of points with the discrete topology, we may take the basic open sets to be $U_q = \{v \in B(X) \mid q \in v\}$. If K is finite, $\bigcap_{q \in K} U_q$ is $\{v \in B(X) \mid K \subset v\} = O_K$. Every open set of B(X) contains an open set of this type.

Suppose X is dense in Y. Any map from X to any Z has at most one extension to Y. It follows that B(Y) → B(X) is 1-1.

2. Elementary Results about Continuous Colorings

We will show that B applied to a disjoint union is a product, and if the two structures are connected, B applied to a product is a disjoint union. We apply a theorem of Lovasz and Greenwell to show that for simplicial complexes, the colorings of an infinite product are given by colorings of one of the factors. Let I = (T,V) and J = (S,W) be two G-structures. The product I×J is the G-structure with underlying topological space V×W. For each t in T, s in S, and g in G we have a block (t,s,g). If $x \in \Omega$ then (t,s,g) applied to x is the point (t(x),s(g(x))) of V×W.

In order to understand why the factor g appears, suppose that I and J have no vertices in common. A coloring F of I ∐ J consists of a colorings h and k of I and J. However, if g ∈ G, then h and gk is also a coloring. If g is not the identity, this coloring is different from F. We thus see that B carries disjoint union into a product.

We define a G-structure (T,V) to be connected if any two points x and y of V are joined by a chain of adjacent points. If Ω is connected, this is is equivalent to saying that V is connected.

Examples

(1) Consider the G-structure Ω_G. $\Omega_G \times \Omega_G$ has point set Ω×Ω, and one block for each element of G. A block has the form $\{(x,gx) \mid x \in \Omega\}$ for g ∈ G.

(2) Let Ω be the real numbers, and let G be the reals acting by addition. g(x) = g+x. $\Omega_G \times \Omega_G$ has underlying space $\mathbf{R}\times\mathbf{R}$ and the blocks $\{(x,x+g) \mid g\in\mathbf{R}\}$ are all lines of slope 1 .

(3) Let Ω be the reals again, and let G be all homeomorphisms of **R**. The blocks of $\Omega_G \times \Omega_G$ are the graphs of homeomorphisms.

$\Omega_G \times \Omega_G$ is not connected in (2), but is in (3). The difference is that the group of (2) is too small.

Lemma 1

If G is 2-transitive ,then $\Omega_G \times \Omega_G$ is connected.

Proof:

This is the same as for the purely group case of Chapter 5. □

Corollary 2

If G is 2-transitive, I a connected G structure, then I×I is connected

Proposition 3

If G is 2-transitive, I and J connected G-structures, then B(I×J;G) = B(I;G) $\coprod$ B(J;G).

Our next goal is to extend Proposition 3 to infinite products. We are able to do this only in the case Ω is finite with the discrete topology. The key result is found in [Lovasz, Greenwell ,1974]. If A is a set, define an 0-1 measure on A to be a function μ on all subsets of A which satisfies

(1) $\mu(B)$=0 or 1 for all subsets B of A.

(2) $\mu(B\cup C) = \mu(B)+\mu(C)$ for all disjoint pairs B,C of subsets of A.

(3) $\mu(A) = 1$

The measure is trivial if there is a point a in A such that $\mu(\{a\}) = 1$. From any 0-1 measure μ and (n-1)-complexes X_i, there are colorings of $\prod_{i \in A} X_i$. Let f_i be a coloring of X_i and let $(x_i) = x$ be in $\prod_{i \in A} X_i$. Let $S_c = \{a \in S \mid f_a(x_a) = c\}$. Then the $S_1, \cdots S_n$ are disjoint and so partition A. There is exactly one k such that $\mu(S_k) = 1$. Define $f_\mu(x) = k$. f_μ is a coloring of $\prod_{i \in A} X_i$.

Theorem (Lovasz, Greenwell, and Lempert)

All colorings of $\prod_{i \in A} X_i$ are induced by colorings of the factors and a 0-1 measure of A.

Theorem 4

If all X_i are connected simplicial complexes, then $B(\prod_{i \in A} X_i) = \coprod B(X_i)$ where the colorings of the left hand side are continuous.

Proof:

As in the proof for two factors, it suffices to show the theorem holds for the special case of where each X_i is Δ^{n-1}. A basis for the topology on $\prod_{i \in A} X_i$ is given by sets of the form $x_1 \times x_2 \times \cdots x_k \times \Delta^{n-1} \times \Delta^{n-1} \times \cdots$ for some k.

By the Theorem, we may assume the coloring is given by a 0-1 measure μ. We will show that if μ is not trivial ,then f_μ is not continuous. If μ is not trivial, then μ is zero on all finite sets. It follows that if two vertices of $\prod_{i \in A} X_i$ differ in a finite number of coordinates, then they are colored the same by f_μ. If we choose any basic set O of the above product form, then it is clear that for any vertex x of $\prod_{i \in A} X_i$ there is a y in O which differs from x in a finite number of coordinates. It follows that every open set contains a point colored the same as x. This implies that every basic open set of the form $\{v \in B(\prod_{i \in A} X_i) \mid v \in p\}$ contains a point colored the same as x. This is impossible, since X is arbitrary. □

3. Infinite Reflexive Complexes

In this section we prove that various infinite complexes are reflexive, including $\hat{\mathbb{Z}}$, $\Delta^\infty \# \Delta^\infty$, and certain infinite binary trees.

Theorem 1

Suppose that $X_1 \subset X_2 \cdots$ is an infinite sequence of complexes and $X = \bigcup X_i$. If

(1) Each X_i is a retraction of X_{i+1}

(2) Every independent set of X_i is contained in a vertex of $B(X_i)$

(3) Each X_i is reflexive

then X is reflexive.

Proof:

From (1), there are maps $J_k : X_{k+1} \to X_k$ such that J_k is the identity on X_k. Since each J_k is onto, B(J_k) is 1-1. Let $B_*(X) = \bigcup B(J_k)$.

It is not true that $B_*(X) = B(X)$, but we will show that $B_*(X)$ is dense in B(X) and B($B_*(X)$) = X. It follows that since $B_*(X)$ is dense in B(X), and every coloring of $B_*(X)$ extends to B(X), X = B($B_*(X)$) = B(B(X)).

To show that $B_*(X)$ is dense in B(X), let O be a basic open set of X. Since the underlying topological space of Ω is a finite set of points with the discrete topology , O = $\{f^{-1}(p) \mid C \subset f^{-1}(p)\}$ where C is a finite independent subset of X. If we choose k large enough, C is contained in X_k. By (2), there is a vertex of B(X_k) containing C. If the coloring containing C is f, then by using the retraction maps, we get colorings of all the X_i which agree on intersections. Thus we get a coloring of X which is in $B_*(X) \cap O$, so $B_*(X)$ is dense in B(X).

Let f be a coloring of $B_*(X)$. f restricts to a coloring of each $B(X_i)$. Since X_i is reflexive, f| X_i is determined by some top simplex of X_i. A vertex of f corresponds to a sequence $\{v_i\}$ where $v_i \in X_i$. We show that if the sequence $\{v_i\}$ is not eventually constant, then any coloring containing the vertex is not a continuous map, and so is not in B(X). Assume $f^{-1}(p)$ corresponds to $\{p_i\}$. If f is continuous, then $f^{-1}(p)$ is open, so there is a finite set of vertices C of X such that O = $\{g^{-1}(p) | \ g \in B(X)\ ,\ g(C)=p\}$ is contained in $f^{-1}(p)$. Suppose that $p_k \notin C$ for all large enough k. By choosing an independent set containing v_k and a point adjacent to p_i, then we can find a coloring h of X such that $h(k) = p \ \neq h(v_i)$. This shows that h is in O, yet not in $f^{-1}(p)$. This shows that $\{v_i\}$ is eventually constant, and hence B($B_*(X)$) = X. □

Corollary 2

The following are reflexive

(1) $\hat{\mathbb{Z}}$

(2) $\hat{\mathbb{Z}}^+$

(3) Any complex formed as follows : begin with Δ^n. To any complex already formed, you may add Δ^n to any codimension 2 face which previously lay in exactly one top simplex.

Proof:

We must verify that the conditions of the theorem are met. (1) is clear. In (2), $\hat{\mathbb{Z}}^+$ is the hatted version of the half- infinite line. For (3), the finite version follows from Chapter 6. □

Let Δ^∞ be the integers with the discrete topology. We think of Δ^∞ as the infinite simplex. In order to generalize the result that $\Delta^n \# \Delta^n$ is reflexive, the topology on Δ^∞ is necessary.

Theorem 3

$\Delta^\infty \# \Delta^\infty$ is reflexive

Proof:

The idea behind the argument is the same as the finite case, but there are some topological complications. Consider a vertex x of $B(\Delta^\infty \# \Delta^\infty)$ to be an infinite permutation of the integers. Let [x] denote the set of all permutations which disagree with x in only a finite number of places.

Since Δ^∞ has the discrete topology, the open sets of $B(\Delta^\infty \# \Delta^\infty)$ are generated by the open sets O_K consisting of all permutations which have the same values in a finite set K of positions. We first observe that if O is any non-empty open set, then $[x] \cap O$ is non-empty for any permutation x. Indeed, choose some O_K contained in O, and construct a permutation which has the prescribed values on K and disagrees from x only on a finite set.

Next, if f is a coloring of $B(\Delta^\infty \# \Delta^\infty)$, $f^{-1}(p)$ is a vertex of $B^2(\Delta^\infty \# \Delta^\infty)$. For all x, we can not have that [x] is contained in $f^{-1}(p)$, for then $[x] \cap f^{-1}(q) = \emptyset$ for $p \neq q$, yet $f^{-1}(q)$ is an open set.

Let Y be a vertex of $B^2(\Delta^\infty \# \Delta^\infty)$. Suppose that $A \in Y$ is a permutation. If we transpose two elements of A, the resulting permutation might or might not belong to Y. We can not have that all transpositions applied to A lie in Y, for the set of all such permutations determined by transpositions applied repeatedly to A is [A]. So let A be a permutation of Y, and B a permutation resulting from A by a single transposition which is not in Y. We may assume that they are as in Figure 1.

A	... -1 0 1 2 3 4 5 6 7 8 ...
B	... -1 0 2 1 3 4 5 6 7 8 ...
C	... 0 2 1 3 4 5 6 7 8 9 ...
W(5,3)	... -1 0 1 3 4 5 2 6 7 8 ...

Figure 1

Since any partial infinite Latin square can be extended to a Latin square, the Extension property holds for $B(\Delta^{\infty}\#\Delta^{\infty})$. In other words, give any two permutations C and D such that they have different values in every position, there is a coloring of $\Delta^{\infty}\#\Delta^{\infty}$ with C and D as vertices. Consequently, we conclude that if C and D are both in Y, they must agree is some position. In order to prove the theorem, it suffices to show that they agree in the same position.

If X is a permutation, let nX denote the result of shifting the permutation n places to the right. 0X = X and {nX | n an integer} is a coloring of $\Delta^{\infty}\#\Delta^{\infty}$. Consequently, for every permutation X, there is a unique n such that nX is in Y. The only n for which nB meets A are -1,0,1. Since 0B is not in Y, either 1B or -1B are in Y. By symmetry, assume that -1B is in Y and let -1B = C. We will eventually show that Y consists of all permutations with 1 in the same position (call it the first position) as A.

For $2 \leqslant k \leqslant n$, define the permutation

$$\mathrm{W(n,k)} = \cdots -1,0,1,k,k+1, \cdots n,2,3,4, \cdots k-1,n+1,n+2,n+3, \cdots$$

W(5,3) is shown in Figure 1. The only m for which mW(n,k) intersect A are 0, k-2, and k-n-1. The m for which mW(n,k) intersects C are 0, -1, k-3, k-n-3. As only W(n,k) meets both A and C, W(n,k) is in Y.

We can now conclude that if σ is any permutation of [1,n] such that the permutation $\cdots,-2,-1,0,\sigma,n+1,n+2,\cdots$ = V is in Y , then 1 is fixed by σ. There are n-1 permutations W(n,k) for k=2, ... n and if V meets each one of them and does not have 1 in the first position of σ, there are only n-2 positions for V to meet the n-1 W's. Similarly, we can argue that if τ is any permutation of [-m,0], then $\cdots -(m+1),\tau,1,\sigma,n+1, \cdots$ is in Y.

Next, assume that D is a permutation in Y that meets A in a finite number of places. Write D = GσH, where σ meets A, but G and H are disjoint from A. If 1 is not in the first position of D, we can find permutations τ and σ such that $\cdots -(m+1),\tau,1,\sigma,(n+1), \cdots$ is disjoint from D yet is in Y. Consequently, 1 is in the first position of A.

Suppose that D is any permutation which meets A in a finite number of places. There is some member E of [D] such that E is in Y, because [D] $\cap$Y is non-empty. We can follow the previous argument and conclude that [D]$\cap$Y consists of all permutations in [D] taking a value x in the j-th position. But, from the previous paragraph, they all have a 1 in the first position.

Finally, let E be any permutation in Y. Find a permutation D which meets E and A in a finite number of places. The above paragraph shows that E has a 1 in the first position, and so we are done . □

4. Cartesian Products and Latin Square Spaces

In this section we define the cartesian product for G-structures. Using this product, we then define Latin Square spaces. In particular, S^3 is a Latin Square space, but S^2 is not.

Let I = (T,V) and J = (S,W) be two G-structures. I#J has underlying space V×W with the product topology. The structure maps are the same as for the group structure of Chapter 5. Define Homeo(X) to be the group of all homeomorphisms of X. Let Ω be the reals **R**. As an Homeo(**R**)-structure, **R**#**R** has the plane as its topological space ; the blocks are all vertical and horizontal lines.

It is not easy to create colorings of I#J. The only known way is for Ω to have some extra structure.

Theorem 1

If Ω is a topological group, I and J are G-structures where $\Omega \subset G$, then I and J have colorings iff I#J does.

Proof:

The map $I \to I\#J$ given by sending x to (x,y) for a fixed y is continuous. Consequently, if I#J has a coloring, so do I and J. Conversely, let the group operation be $\circ$. If f is a coloring of I and g is a coloring of J, define $fg : I\#J \to \Omega$ by $fg(x,y) = f(x) \circ g(y)$. Since $\circ$ is continuous, fg is also. For any fixed x, the map fg restricted to x#W is a homeomorphism, so fg is a coloring . □

Recall that a Latin square is equivalent to a map $\Delta^n \# \Delta^n \to \Delta^n$. We now extend the definition to arbitrary spaces. Ω is a Latin square space iff as a Homeo(Ω) structure, $\Omega \# \Omega$ has a coloring. We see that if G is a topological group, then G is a Latin square space.

Suppose that Ω is a Latin square space and F is a coloring of $\Omega \# \Omega$. For any x in Ω, $F|\, x\#\Omega$ is a homeomorphism, so for each y in Ω there is a unique $F^t(x,y)$ such that $F(x,F^t(x,y)) = y$. We claim that F^t is also a coloring. Since $F(x,\bullet)$ is a homeomorphism, and $F(x,F^t(x,F(x,y))) = F(x,y)$, we have $F^t(x,F(x,y)) = y$. Equivalently, $F(a,p)=b$ iff $F^t(a,b)=p$. This shows that F^t is a bijection if either coordinate is fixed.

To show F^t continuous, let $F^t(x,y) = p \in O$, where O is an open set of Ω. Let O_2 be an open set containing y. Since F is continuous, there are open sets O_1 containing x and O_3 containing p such that $F(O_1, O_3) \subset O_2$. Since $F(O_1, O_3 \cap O) \subset O_2$, it follows that $F^t(O_1, \mathrm{Im}(F)) \subset O_3 \cap O$. But $F(x,\bullet)$ is a homeomorphism, so Im(F) contains an open set O' containing y. $F^t(O_1, O') \subset O$, $x \in O_1$, $y \in O'$, and so F^t is continuous.

Theorem 2

Let Ω be a topological group. The vertices of $B(\Omega \# \Omega)$ are in 1-1 correspondence with Homeo(Ω).

Proof:

If f is a homeomorphism of Ω, $F(x,y) = f(x) \circ y$ is a coloring. Conversely, let F be a coloring and $p \in \Omega$. Since $F(x,F^t(x,p))=p$, $F^{-1}(p) = \{(x,F^t(x,p)) \mid x \in \Omega\}$. F^t is a coloring, hence $F^{-1}(p)$ is the graph of the homeomorphism $F^t(\bullet,p)$. □

We will now speak of homeomorphisms of Ω instead of vertices of $B(\Omega \# \Omega)$.

Theorem 3

If Ω is compact, connected, f,g $\in$Homeo(Ω), and f and g are connected in $B(\Omega \# \Omega)$, then f is isotopic to g.

Proof:

It suffices to show that if F is a coloring of $\Omega \# \Omega$, and $p,q \in \Omega$, then the homeomorphism corresponding to $F^{-1}(p)$ is isotopic to the one corresponding to $F^{-1}(q)$. Let $\gamma:[0,1] \to \Omega$ start at p and end at q. Define $H : \Omega \times [0,1] \to \Omega$ by $H(x,s) = F^t(x,\gamma(s))$. H is continuous, and for fixed s, H(x,s) is a homeomorphism. H is the desired isotopy, for H(x,0) corresponds to $F^{-1}(p)$ and H(x,1) to $F^{-1}(q)$. □

Each component of $B(\Omega \# \Omega)$ determines a single isotopy class of Ω. Since a Lie group has more than one isotopy class of homeomorphisms, it follows that $B(\Omega \# \Omega)$ is not connected if Ω is a Lie group.

We next claim that the connected components of $B(\Omega \# \Omega)$ are homeomorphic. Aut($\Omega \# \Omega$) acts on $B(\Omega \# \Omega)$. Choose any f,g in Homeo(Ω). $H : (x,y) \to (x,gf^{-1}(y))$ is an automorphism of $B(\Omega \# \Omega)$. H carries the vertex $\{(x,f(x)) | x \in \Omega\}$ which corresponds to f to the vertex $\{(x,g(x)) | x \in \Omega\}$ corresponding to g. Consequently, Aut($\Omega \# \Omega$) is transitive on $B(\Omega \# \Omega)$.

Finally, we ask which spaces are Latin square spaces. Let F be a coloring of $\Omega\#\Omega$. Choose a base point $*$ in Ω and set $F|*\#\Omega = h_1$ and $F|\Omega\#* = h_2$. The map $F(h_1^{-1}h_2^{-1})$ makes Ω into an H space. Since the two sphere is not an H-space, and the three sphere is, we have

Theorem 4

(1) If Ω is a Latin square space, then Ω is an H-space.

(2) The two-sphere is not a Latin square space.

(3) The three sphere is a Latin square space.

5. Colorings of Real Latin Squares

In this section we study the structure of $B(\mathbf{R}\#\mathbf{R})$ in detail. We show that $B(\mathbf{R}\#\mathbf{R})$ has exactly two connected components, and that the automorphisms of $\mathbf{R}\#\mathbf{R}$ inject into the automorphisms of $B(\mathbf{R}\#\mathbf{R})$. Using a finite compactly supported measure, we construct colorings of $B(\mathbf{R}\#\mathbf{R})$. As opposed to the case of coloring with finitely many colors, it is difficult to tell that two colorings are not equivalent. We are able to show that many kinds of colorings are distinct, thus showing that $B(\mathbf{R}\#\mathbf{R})$ is not close to being reflexive.

From the last section, the vertices of $B(\mathbf{R}\#\mathbf{R})$ correspond to homeomorphisms of $\mathbf{R}$. The basic open sets of $B(\mathbf{R}\#\mathbf{R})$ are of the form $\{g\in\text{homeo}(\mathbf{R})| \mid g(x)-f(x)| <\epsilon\}$, $\{g^{-1}\in\text{homeo}(\mathbf{R})| \mid g(x)-f(x)| <\epsilon\}$, and $\{g\in\text{homeo}(\mathbf{R})| \mid g(x)-f(x)<\epsilon \text{ for all } x\in\mathbf{R}\}$.

Lemma

If f and g $\in$homeo($\mathbf{R}$), then f and g lie in the same coloring iff their graphs are disjoint.

Proof:

Assume for all x, $f(x) < g(x)$. For $0\leqslant\lambda\leqslant1$, let $F(x,\lambda g(x)+(1-\lambda)f(x)) = \lambda$. This defines F(x,y) for all (x,y) such that $f(x) \leqslant y \leqslant g(x)$. If $y \geqslant g(x)$, let F(x,y) = y-g(x)+1. If $y \leqslant f(x)$, let F(x,y) = y-f(x). F is seen to be a coloring of **R**#**R**. Since $F^{-1}(0)$ corresponds to f and $F^{-1}(1)$ corresponds to g, we are done . □

Theorem 1

B(**R**#**R**) has exactly two connected components

Proof:

Let f and g be two orientation preserving homeomorphisms, and let h(x) = max(f(x),g(x)). h is a homeomorphism of **R**. h+1 is disjoint from both f and g, so by the Lemma, h+1 is adjacent to f and g. It follows that the two components consist of the orientation preserving and the orientation reversing homeomorphisms. □

Theorem 2

Aut(**R**#**R**) ≅ **R**#**R**#Z_2

Proof:

It suffices to show that if h is a homeomorphism of **R**#**R** which has the property that h carries the graph of every homeomorphism of **R** to the graph of a homeomorphism of **R**, then h(x,y) = (f(x),g(x)) or (g(y),f(x)), where f,g∈homeo(**R**). Let L_x and L_y be the lines parallel to the axis through (x,y), and let L_x' and L_y' be the lines through h(x,y) parallel to the axis. The four quadrants determined by removing L_x and L_y map to the quadrants determined by removing L_x' and L_y'. Since h is a homeomorphism, we may assume $h(L_x) = L_x'$ and $h(L_y) = L_y'$.

We conclude that h carries vertical lines to vertical and horizontal to horizontal. Let f(x) = h(x,0) and let g(y) = h(0,y). To find h(x,y), we note that (x,y) is at the intersection L_x and L_y. Consequently, h(x,y) is the intersection of the horizontal line through (0,g(y)) and the vertical line through (f(x),0), which is (f(x),g(y)) . □

Theorem 3

$\mathrm{Aut}(\mathbf{R}\#\mathbf{R}) \to \mathrm{Aut}(\mathrm{B}(\mathbf{R}\#\mathbf{R}))$ is injective.

Proof:

A member α of the kernel fixes every vertex of $\mathbf{R}\#\mathbf{R}$. Since all vertices of a coloring have the same orientation, a vertex V(h) of $\mathrm{B}(\mathbf{R}\#\mathbf{R})$ is of the form $\{(x,h(x)) \mid x \in \mathbf{R}\}$, where $h \in \mathrm{homeo}(\mathbf{R})$. α applied to V(h) is $\{(f(x),g(h(x)) \mid x \in \mathbf{R}\} = \mathrm{V}(\mathrm{ghf}^{-1})$. Consequently, for all $h \in \mathrm{homeo}(\mathbf{R})$, $\mathrm{h} = \mathrm{g} \circ \mathrm{h} \circ \mathrm{f}^{-1}$. Setting h = f shows that g=f and so $\mathrm{h} \circ \mathrm{f} = \mathrm{f} \circ \mathrm{h}$. If we let h(x)=x+t, then f(x)+t=f(x+t). Setting x=0 gives f(t)=t+f(0). If we choose $\mathrm{h(x)=x^3}$, then $(\mathrm{x+h(0)})^3 = \mathrm{f(0)+x^3}$, hence f(x)=x. Since f and g are the identity, α is the identity. $\square$

For the remainder of this section we will investigate the colorings of the connected component of $\mathrm{B}(\mathbf{R}\#\mathbf{R})$ consisting of the orientation preserving homeomorphisms. The next result shows how to construct a new class of colorings.

Theorem 4

If μ is a finite compactly-supported Borel measure on $\mathbf{R}$, $\overline{\mu}(f) = \int f \cdot d\mu$ is a coloring of $\mathrm{B}(\mathbf{R}\#\mathbf{R})$

Proof:

Since μ is finite with compact support, $\overline{\mu}(f)$ is well-defined. If a<b, then $\overline{\mu}(f)^{-1}(a,b) = \{f \mid \int f \cdot d\mu \in (a,b)\}$. For each f in $\overline{\mu}(f)^{-1}(a,b)$, choose an ϵ such that $\int f \cdot d\mu \pm \epsilon \in (a,b)$. If $O = \{g \mid |f(x)-g(x)| < \epsilon\}$, then for $g \in O$, $|\int g \cdot d\mu - \int f \cdot d\mu| \leqslant |\int (g-f) d\mu| \leqslant \epsilon$, so $O \subset \overline{\mu}(f)^{-1}(a,b)$. This shows that $\overline{\mu}$ is continuous.

$\overline{\mu}$ restricted to a coloring of $\mathrm{B}(\mathbf{R}\#\mathbf{R})$ maps homeomorphically to $\mathbf{R}$. If $f(x) > g(x)$ for all x, then $\int f \cdot d\mu > \int g \cdot d\mu$. As f gets large (resp. small) $\overline{\mu}(f)$ gets arbitrarily large (resp. small) on the support of μ, so $\overline{\mu}(f)$ goes to ∞ (resp. $-\infty$) as f gets large (resp.

small). Consequently, $\overline{\mu}$ restricted to a coloring is a homeomorphism. □

If $\alpha \in \mathrm{Aut}(\mathrm{B}(\mathbf{R}\#\mathbf{R}))$, then $\int \alpha(f) \cdot d\mu$ is also a coloring. Consider the case where μ is the point measure with all its mass at p. $\overline{\mu}(f) = \int f \cdot d\mu = f(p)$, so $\overline{\mu}$ applied to f is the value of f where its graph intersects the line x=p. This corresponds to the coloring of B(R#R) determined by the line x=p under Φ. Similarly, for the coloring $\overline{\mu}(f) = \int f^{-1} \cdot d\mu$, the value is $f^{-1}(p)$, so this coloring corresponds to the line y = p under Φ. If **R**#**R** were reflexive, these would be all the colorings.

If $g \in \mathrm{homeo}(\mathrm{R})$ and μ is a point measure with its mass at p, then $\overline{\mu_g}(f) = \int gh \cdot d\mu = gh(p)$. Consequently, $\overline{\mu_g} = g\overline{\mu}$ and so $\overline{\mu}$ and $\overline{\mu_g}$ are equivalent colorings.

Lemma 5

If μ and τ are finite compactly supported Borel measures and the colorings that they determine have a vertex in common, then $\mu = c\tau$ for some constant c and the colorings are equivalent.

Proof:

We may assume that $\int 1 \cdot d\mu = \int 1 \cdot d\tau = 1$. We show that for all $f \in \mathrm{homeo}(\mathbf{R})$, { $\int f \cdot d\mu = p$ iff $\int f \cdot d\tau = q$ } implies $\mu = \tau$. Suppose $\int f \cdot d\mu = p$. For any λ, $\lambda f + (1-\lambda)p$ $\in \mathrm{homeo}(\mathbf{R})$ and $\int (\lambda f + (1-\lambda)p) \cdot d\mu = p$, so $\int (\lambda f + (1-\lambda)p) \cdot d\mu = q$. Thus, $\lambda \int f \cdot d\tau + (1-\lambda)p = q$, and so $\int f \cdot d\tau = (q-(1-\lambda)p/\lambda$ for all λ. As $\lambda \to \infty$, we see that $(q-p)/\lambda + p \to p$, so p=q.

If f is any function in homeo(**R**), let I $= \int f \cdot d\mu$. $\int (f-I) \cdot d\mu = p$ so $\int (f-I) \cdot d\tau = p$. Thus for all $f \in \mathrm{homeo}(\mathbf{R})$, $\int f \cdot d\mu = \int f \cdot d\tau$. This easily implies that $\mu = \tau$. □

Theorem 6

Suppose μ and τ are finite compactly supported Borel measures. If there are real numbers p and q such that

$$\int f\bullet d\mu = p \text{ iff } \int f^{-1}\bullet d\tau = q$$

then μ is a point measure with mass at q and τ is a point measure with mass at p.

Proof:

We may assume $\int 1\bullet d\mu = 1 = \int 1\bullet d\tau$. Let $\int f\bullet d\mu = p$ and let C be a compact set of **R** containing the support of τ. For any real number t, define $g_t(x) = tf(x)+(1-t)p$. $\int g_t\bullet d\mu = t\int f\bullet d\mu + \int (1-t)p\bullet d\mu = tp+(1-t)p = p$, and hence $\int g_t\bullet d\tau = q$.

Since $g_t^{-1} = f^{-1}(p+(x-p)/t)$, $\int f^{-1}(p+(x-p)/t)\bullet d\tau = q$ for all t. For any $\epsilon > 0$ we choose a t so large that $|f^{-1}(p+(x-p)/t)-f^{-1}(p)| < \epsilon$ for all x in C. Integrating the inequality

$$f^{-1}(p)-\epsilon \leqslant f^{-1}(p+(x-p)/t) \leqslant f^{-1}(p)+\epsilon$$

and using the fact that τ is zero outside C, we get

$$f^{-1}(p)-\epsilon \leqslant \int f^{-1}(p+(x-p)/t)\bullet d\tau \leqslant f^{-1}(p)+\epsilon$$

Consequently, $f^{-1}(p)-\epsilon \leqslant q \leqslant f^{-1}(p)+\epsilon$ and so $f^{-1}(p) = q$ and f(q)=p.

Now let h be an arbitrary homeomorphism of **R**. The integral with respect to μ of $h+p-\int h\bullet d\mu$ is p, so we apply the above to find that $h(q)+p-\int h\bullet d\mu = p$, so $h(q) = \int h\bullet d\mu$. Since this holds for all homeomorphisms h, μ is a point measure with mass at q. Similarly, τ has all its mass at p . □

Corollary 7

If μ and τ are finite compactly supported Borel measures and if the colorings $\int f\bullet d\mu$ and $\int f^{-1}\bullet d\tau$ have a vertex in common then μ and τ are point measures.

Lemma 8

Suppose that μ and τ are finite compactly supported Borel measures such that $\mu[-\infty,x]$ and $\tau[-\infty,x]$ are continuous in x. If $g\in$homeo(**R**) and for all $f\in$homeo(**R**) such that $\int f\cdot d\mu = 0$, we have $\int g(f)\cdot d\tau = 0$ then $g(x) = \gamma x$ and $\mu = \beta\cdot\tau$ for constants γ and β.

Proof:

Since μ is continuous, if h(x) is a non-decreasing step function satisfying $\int h\cdot d\mu = 0$, we may find $f_n\in$homeo(**R**) such that $h = \lim_{n\to\infty} f_n$ and $\int f_n\cdot d\mu = 0$. Since τ is continuous, we conclude that $\int g(h)\cdot d\tau = 0$.

We introduce a particular step function. Assume [a,b] contains the support of τ. For a $\leqslant$ p $<$ q $\leqslant$ b, let h be s on $[-\infty$,p], 0 on (p,q), and t on [q,∞] . Define the distribution functions G(x) $= \tau[-\infty,x]$ and F(x) $= \mu[-\infty,x]$. Normalize μ and τ so that F(∞) = G(∞) = G(b) = F(b) = 1.

Evaluating $\int h\cdot d\mu$ and $\int g(h)\cdot d\tau$ gives

$$sF(p) + t(1-F(q)) = 0 \tag{6}$$

$$g(s)G(p) + g(0)(G(q)-G(p)) + g(t)(1-G(q)) = 0 \tag{7}$$

Solving for s and substituting gives

$$g(-t(1-F(q))/F(p))G(p) + g(0)(G(q)-G(p)) + g(t)(1-G(q)) = 0 \tag{8}$$

If we let q approach p in (8), we find g(0) = 0. We will show that (9) implies g(x)=x and G(x)=F(x). If we had not normalized, we would get the constants γ and β.

$$g(-t(1-F(q))/F(p))G(p) + g(t)(1-G(q)) = 0 \tag{9}$$

Let $x = (1-F(q))/F(p)$. Then $p = F^{-1}((1-F(q))/x)$ and $g(tx) = g(t)(1-G(q))/G(F^{-1}((1-F(q))/x))$. If we denote the second factor by k(x), then g(tx) = g(t)k(x). If we choose q close enough to b, x can take the value 1. Consequently, k(1) = 1. If we define W(v) = g(v)/g(1), then W(tv) = W(t)W(v). Since W is continuous, there are constants c and e such that $W(v) = cv^e$. This implies that $g(v) = cv^e$. We may assume c=1.

This relation and (9) (with p=q) show that $G(p) = \dfrac{1}{1-(F(p)-1)/F(p)^e}$. From (9) with t=1, we get after some manipulation $F(p)^e-(F(p)-1)^e = F(q)^e-(F(q)-1)^e$ which is certainly false unless F is constant or e is 1 □

Corollary 9

If μ and τ are as in Lemma 8, then the colorings $\int h(f)\bullet d\mu$ and $\int g(f)\bullet d\tau$ have no vertex in common in $B^2(\mathbf{R}\#\mathbf{R})$ unless $\mu = b\tau$ and $g(x) = h(x+c)+d$ for constants b,c and d. The colorings are equivalent.

Proof:

If they have a vertex in common, then there are p and q such that $\int h(f)\bullet d\mu = p$ iff $\int g(f)\bullet d\tau = q$. Applying the Lemma to h(f)-p and $g(h^{-1})(x+p)-q$ gives the result . □

We have seen that $B^2(\mathbf{R}\#\mathbf{R}) = \mathbf{R}\#\mathbf{R} \coprod X$, where X is non-empty. We conjecture that no two colorings of X share a vertex. Lemma 5, Theorem 6, Corollary 7, and Corollary 9 all support this. It may be that $(\mathbf{R}\#\mathbf{R})$ is reflexive.

Coloring with Arbitrary Complexes

In this Chapter we introduce a generalization of coloring that will allow us to develop a theory of coloring polytopes built out of regular polyhedra that is entirely parallel to the theory of coloring complexes constructed from simplices. The first section contains the basic results on the category, the Hom functor, and the coloring functor. The second section develops the theory of coloring surfaces which are composed of squares. The results are remarkably close to those for the usual 4-coloring [Fisk,1978]. Next, we look at decompositions based on pentagons. Lastly, we generalize to higher dimensions.

1. Introduction

Let **Z** be a fixed simplicial complex, and let G be a group of automorphisms of **Z**. If G is not specified, we take G to be all automorphisms of **Z**. A **Z**-complex is a simplicial complex X, and a collection $\{f_i\}$ of homeomorphisms of **Z** into X. All simplices of X are contained on the image of some f_i. A **Z**-map from one **Z**-complex to another is defined the the same as in Chapter 6. We will show that B, Hom, and Aut are functors in the category of **Z**-complexes and **Z**-maps. We first give some examples.

(1) Let **Z** be Δ^n, and G any subgroup of Sym(n+1). This is the set-up of Chapter 6.

(2) Let **Z** be P_4. Take the 64 squares of a chessboard, and map P_4 to each square. This is a **Z**-complex. There are decompositions into squares of all 2-dimensional surfaces. In the next section we study their colorings in detail.

(3) Let **Z** be I^k, the k-cube. The theory of **Z**-coloring decompositions of an k-manifold into k-cubes parallels the simplicial theory.

(4) Let **Z** be the octahedron. The 24-cell [Coxeter,1948], a regular polytope of 4 dimensions, is a **Z**-complex. It is not hard to see that there are no **Z**-maps from the 24-cell to the octahedron.

(5) Let **Z** be any complex with n vertices. Make Δ^{n-1} into a **Z**-complex by taking the structure maps to be the n! different embeddings of **Z** into Δ^{n-1}.

Since the category of **Z**-complexes is so similar to the categories of the previous two chapters, we will not go into complete detail. Define $\mathrm{Hom}_{\mathbf{Z}}(X,Y)$ to have vertices all **Z**-maps from X to Y. A structure map $F:\mathbf{Z} \to \mathrm{Hom}_{\mathbf{Z}}(X,Y)$ satisfies the condition: If u is adjacent to v in **Z**, then the maps f(u) and f(v) are adjacent in Hom(X,Y). If we define cartesian product as before, then the evaluation map $\mathrm{Hom}_{\mathbf{Z}}(X,Y)\#X \to Y$ is a **Z**-map.

A **Z**-coloring of X is a **Z**-map from X to **Z**. A **Z**-coloring f determines vertices $f^{-1}(z)$ of $B_{\mathbf{Z}}(X)$, for $z \in Z$. The structure map *fStar* of $B_{\mathbf{Z}}(X)$ is defined by $fStar(z) = f^{-1}(z)$.

We define Φ: $X \to B^2_{\mathbf{Z}}(X)$ by sending x in X to $\{v \in B_{\mathbf{Z}}(X) | x \in v\}$. To show that Φ is a **Z**-map, it suffices to show that $\Phi \circ F$ is a structure map for $B^2_{\mathbf{Z}}(X)$, for each structure map F of X. Define the map $\alpha: B_{\mathbf{Z}}(X) \to \mathbf{Z}$ by setting $\alpha(w)$ to the unique z such that $F(z) \in w$. If α is a **Z**-map, then F is a structure map. Choose a structure map H: $\mathbf{Z} \to B_{\mathbf{Z}}(X)$ given by $H(z) = h^{-1}(z)$ for some **Z**-coloring h of X. $\alpha \circ H(z) = \alpha(h^{-1}(z)) = t$, where $F(t) \in h^{-1}(z)$. This is equivalent to $h(F)(t) = z$. Since both h and F are **Z**-maps, sending z to t is also a **Z**-map.

Examples

(1) Let **Z** be I^3, the 3-cube, and let X be two 3-cubes joined at an edge. An easy calculation shows $B_{\mathbf{Z}}(X)$ consists of two 3-cubes, joined at two edges. Each edge of the intersection is antipodal to the other edge of the intersection. $B_{\mathbf{Z}}(B_{\mathbf{Z}}(X)) = B_{\mathbf{Z}}(X)$, so $B_{\mathbf{Z}}(X)$ is self-dual.

(2) Let **Z** be any complex, and take G the identity. If X is a **Z**-complex, then there is a unique map from **Z** to the image of each structure map. It follows that there is at most one coloring, so $B_{\mathbf{Z}}(X)$ is either **Z** or $\emptyset$.

(3) More generally, if **Z** is any complex, and the stabilizer in G of every point of **Z** is the identity, then for any **Z**-complex X, $B_{\mathbf{Z}}(X)$ is either **Z** or $\emptyset$.

(4) Let **Z** be a tree with three vertices and two edges. If X is a star with 2n edges, take the structure maps to be determined by n disjoint copies of **Z** in X. There are 2^{n-1} distinct colorings, and $B_{\mathbf{Z}}(X)$ is a star with 2^n edges. In case n is 2, X is self-dual.

(5) If **Z** is arbitrary with n vertices, we saw Δ^{n-1} has a natural **Z**-structure. Consequently, S_n has a **Z**-structure in addition to the usual Sym(n) structure. It is easy to see that $\mathrm{Hom}_{\mathbf{Z}}(\mathbf{Z},\Delta^{n-1}) \cong S_n$ as **Z**-structures.

2. Cubical Coloring

In [Fisk,1978] we studied properties of triangulations of the 2-sphere and other 2-manifolds. We show here that if we use squares as the basic unit instead of triangles, we get a similar theory.

We begin with the category C of all graphs whose edge set is the union of 4-cycles. U is $P_4 \cong I^2$, and G is all automorphisms of the square. We will be interested in coloring cubical decompositions of 2-manifolds. These are graphs embedded in 2-manifolds in such a way that all faces are squares. Such graphs are given the obvious U-structure. The difficulty with this category is that if M is a cubical decomposition, then M has at most one I^2 coloring. This is a consequence of the fact that a map from I^2 to I^2 is completely determined by the mapping on one edge. The analogous problem holds for triangulations. We got around it by introducing the hat operation, and studying maps to Δ^3 rather than Δ^2.

A cubical decomposition M has a I^2-structure. Introduce a hat operation by defining $\hat{M}$ to be the result of adding a cube I^3 to each square of M. $\hat{M}$ has an I^3 structure. Any map $M \to I^3$ has a unique extension $\hat{M} \to I^3$. We say that M has a coloring if there is a map from M to ∂I^3. We can now ask questions about cubical decomposition in the spirit of triangulations:

- Does every cubical decomposition of the 2-sphere have a coloring?
- What are even cubical decompositions?
- What are the properties of the degree of a coloring?
- What can one say about $B(\hat{M})$?
- What are the various kinds of local colorings induced by the composition series for the automorphism group of I^3?
- What are the homotopy and cobordism properties of cubical decompositions?

Suppose that M is a cubical decomposition on a surface S. If x is a vertex of M, the degree of x, written $\rho(x)$, is the number of squares of M containing x. If $\rho(x)$ is even, we say x is even and if every interior vertex of M is even, we say M is even. If a vertex is not even, it is odd. Let O(M) be the set of odd vertices of M. If M has a map to I^2, then M is even. Conversely, if M is even and simply connected, then M has a map to I^2. The converse is not necessarily true for non-simply connected surfaces, as is seen by considering the decomposition of the torus into a grid of squares. On the sphere and disk, the underlying graph of a cubical decomposition is bipartite.

Consider a coloring $f:X \to \partial I^3$. Let I^2 have vertices 1,2,3,4; label one face of ∂I^3 with 1,2,3,4, and then label the vertex antipodal to a vertex labeled x with $\overline{x}$. Since ∂I^3 is bipartite, if M has a coloring, then M is bipartite. An edge of X is non-singular if the two squares containing it map to distinct squares. Let ns(f) be the set of all non-singular edges of f. At a vertex, we have an induced map $\text{link}(p) \to P_6$, since the three squares adjacent to a point of I^3 each give two edges to the link. If $\rho(p) = n$, then as each two edges of a square at p map to two distinct edges, we get an induced map $P_n \to P_3$. Consequently, the same arguments with triangulations go through, and we get

Theorem 1

(a) If M is a closed cubical decomposition, then $\partial ns(f) = O(M)$.

(b) If X is even and simply connected, then B(X) is connected.

Proof:

If f is a coloring, and e an edge of ∂I^3, a region D is a Kempe region if $f(\partial D) = e$. There is a unique involution σ of I^3 fixing e. Define the coloring g on D to be f, and off D to be $\sigma \circ f$. We say f is Kempe related to g, and extend this relation to an equivalence relation. The argument analogous to triangulations give that all colorings in (b) are Kempe related, and consequently B(M) is connected. □

There is a nice description of colorings of X in terms of maps to $\partial\Delta^3$. If f is a coloring of M, define $\tilde{f}:M\to\partial\Delta^3$ by sending both x and $\bar{x}$ to x. Since every face of ∂I^3 has four distinct labels (ignoring the bars), we see that $\tilde{f}$ sends the points of a square to distinct points of $\partial\Delta^3$. If we give $\partial\Delta^3$ the I^2-structure of (5), section 1, then $\tilde{f}$ is a valid map. Conversely, if M has such a map, and is bipartite, then we get a coloring f by setting $f = \alpha\times\tilde{f}$ where $\alpha:M \to I^1$ is the bipartite map.

We now investigate degree. If S is a cubical decomposition of a closed surface, a coloring $f:S\to\partial I^3$ is a map between two closed surfaces. Consequently, deg(f) is well defined. If S is oriented, deg(f) is the number of squares mapping to a fixed square of ∂I^3 in an orientation preserving fashion minus those mapping in an orientation reversing way. If S is not oriented, deg(f) is simply the number mod 2 of squares mapping to a fixed square of ∂I^3. Since every point of ∂I^3 meets exactly three faces, the analog of Tutte's Lemma holds:

Lemma 2

Mod 2, deg(f) is the number of odd points of S mapping to a fixed vertex of ∂I^3.

Corollary 3

(a) The degree of any coloring of an even cubical decomposition is even.

(b) If S has less than 8 odd vertices, then the degree of any coloring is even.

We claim that every coloring of an even cubical decomposition of the sphere has degree divisible by 6. To see this, we construct $I^2 \times \partial I^3$. There are 48 squares and 32 vertices. Every edge lies in exactly two faces. The points adjacent to (1,1) are

$$(4,4)(3,3)(2,2)(3,\bar{4})(4,\bar{3})(3,\bar{2})(2,4)(3,3)(4,2)(3,\bar{4})(2,\bar{3})(3,\bar{2})$$

Consequently, $I^2 \times \partial I^3$ is a cubical decomposition of a surface of Euler characteristic 32-96+48=-16. The genus is therefore 9, and every vertex has degree 6. The projection map $I^2 \times \partial I^3 \to \partial I^3$ has degree 6.

If S is a globally even cubical decomposition of a closed surface, then f has a map $f: S \to I^2$. If g is a coloring of S, then we get a map $f \times g$ form S to $I^2 \times \partial I^3$. It follows that deg(f) is divisible by 6. Since any map from the 2-sphere to an orientable surface of non-zero genus has degree zero, the degree of any coloring of an even triangulation of the 2-sphere is zero.

It is easy to see

Theorem 4

(a) If two colorings are Kempe equivalent, then their degrees have the same parity.

(b) If two colorings of a globally even cubical decomposition are Kempe equivalent, then their degrees are congruent mod 12.

For triangulations, each term in the composition series of the automorphism group of Δ^3 correspond to a type of coloring. A similar result holds for cubical decompositions. The composition series is

$$1 \subset \mathbb{Z}_2 \# \mathbb{Z}_2 \subset A(4) \subset S(4) \subset AUT(\partial I^3)$$

A local coloring of a cubical decomposition S is a coherent collection of colorings of st(p) for each p in S. We could define a local coloring of type G to be a coloring such that the map $\pi_1(S) \to \mathrm{Aut}(\partial I^3)$ has its image in G. However, it is more interesting to give extrinsic conditions.

Suppose f is a local coloring such that Im(f) is in S(4). This means all automorphisms of ∂I^3 induced along paths of S preserve the orientation. If F is a face of S, define h(F) to be 1 if the face is positively oriented by f, and , and -1 if negatively. Since f is of type S(4), h is well defined, up to -h.

Conversely, given an assignment h of ± 1 to the faces of S, a necessary and sufficient condition for h to correspond to a local coloring is that the sum of the values of the faces at any point is 0 mod 3. We call such an assignment a Heawood coloring.

If a local coloring is of type A(4), then the automorphisms preserve both the orientation and the bipartite structures. This is a Heawood coloring on a bipartite graph.

Next, suppose a coloring f is of type $\mathbb{Z}_2 \# \mathbb{Z}_2$. f corresponds to a labeling of the edges of a bipartite graph with three labels such that every square has exactly two distinct labels, and they are arranged x,y,x,y.

It is easy to check that an edge coloring locally determines a coloring, as does a Heawood coloring. Perhaps every cubical decomposition of an oriented surface has a local coloring.

Cubical decompositions have homotopy and cobordism theories analogous to those of triangulations. Let S be a cubical decomposition of a surface with boundary such that every interior vertex is even. The boundary of S is the union of circles. A vertex x of the boundary is labeled with $\rho(x)$ mod 2. If S is a cylinder, and the boundary is $A \cup B$, we say the oriented labeled circles are homotopic. If S is arbitrary, we say A and B are cobordant.

As for triangulations, we get a map $\pi_1(S)\rightarrow \mathrm{Aut}(\partial I^3)$. Each of A and B determine the same map, so this is the homotopy invariant. We omit the proof that if two oriented labeled circles have the same homotopy invariant, they are homotopic.

For example, assume that S is a triangulation of the 2-sphere such that all but two vertices are even, and the two odds are adjacent. If F is the square containing the two odd vertices, consider S-F. The boundary of S-F is homotopic to $\emptyset$, so the invariant computed along the boundary should be the identity. However, a simple computation shows that this not so, so such a cubical decomposition does not exist. On the torus, there is such a cubical decomposition, see Figure 1. The two odd vertices are in the lower right corner square.

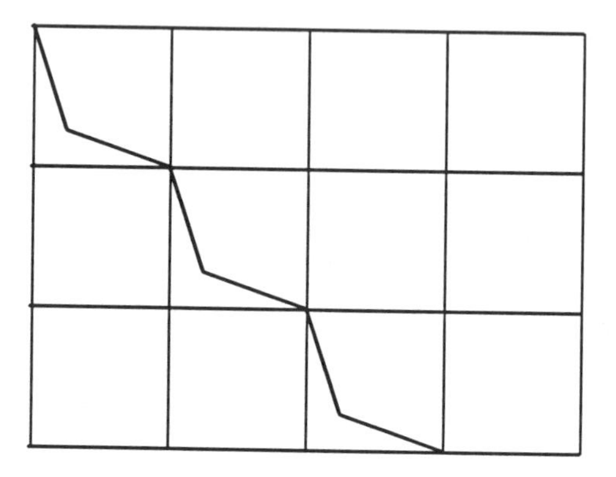

Figure 1

The cobordism invariant is easier. If S is a cobordism between A and B, then the sum of ρ(x) over S equals the sum over the boundary. Since the sum counts every square four times, the sum is zero mod 2. Defining $\rho(A) = \sum_{x\in A} \rho(x)$, we see that ρ(A) is a cobordism invariant.

We now show that $\rho(\bullet)$ is the only invariant. As in the triangulation case, A is cobordant to A and the relation of cobordism is transitive. Let Ω be the free abelian group over $\mathbb{Z}_2$ of all labeled circles modulo A+B, where A and B are cobordant.

Suppose S is a cubical decomposition that induces the correct labeling on A and B, but whose interior contains odd vertices. If $\rho(A) = \rho(B)$, then the number of odd vertices in the interior is even. Choose any two odd vertices in the interior, and consider a path joining them. To one square per edge of the path, add the cubical decomposition of Figure 1 in such a way that the two odd vertices line up on the edge of the path. After these additions, there are two less odd interior vertices. Continuing, we get

Theorem 5

The cobordism group Ω is isomorphic to $\mathbb{Z}_2$.

3. Properties of the Dodecahedron

After having seen the triangle and square, we look at a few of the special properties of the dodecahedron.

The dodecahedron has a P_5 structure. If we try to study P_5 colorings, we run into the same problem: there is at most one coloring of a P_5-decomposition of a surface. We solve the problem by adding a dodecahedron to each pentagon. A coloring of a decomposition of a surface into pentagons is now a map to the dodecahedron.

Just as there is no map from Δ^3 to Δ^2 or I^3 to I^2, it is easy to check that there is no map from the dodecahedron to P_5. If we give Δ^4 the P_5 structure consisting of all 120 embeddings of P_5 in Δ^4, then there are maps from the dodecahedron to Δ^4. Given one such map f, there is a unique map g such that f is disjoint from g, and g is not $\sigma \circ f$ for some permutation σ. Figure 1 shows such a pair.

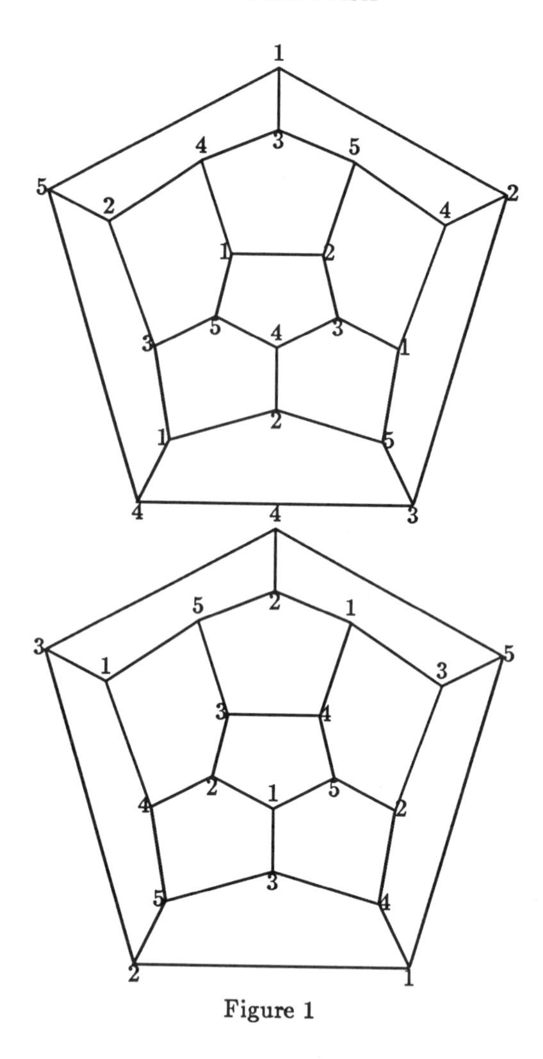

Figure 1

These two mappings have some interesting properties. The faces of each dodecahedron determine cycles of length 5. Twelve of the 24 possible cycles occur in one map and 12 in the other. The vertices of the dodecahedron are the 20 ordered pairs (x,y), $x \neq y$, $x,y \in \mathbb{Z}_5$. The edges at vertex (a,b) are (c,d),(d,e),(e,c) where the permutation (ab)(cde) is in Figure 2.1.5.

4. More Theories

We introduce a class of interesting complexes for which their theories of coloring parallel the triangular theory. These theories are based on cell complexes U which satisfy

(1) U is the union of codimension-1 faces which are all isomorphic to U_1. All codimension-2 faces are isomorphic to U_2.

(2) U is topologically an orientable manifold.

(3) For every codimension-2 face F of U, there is a unique automorphism σ of U whose fixed point set is exactly F. Any automorphism fixing F is either σ or the identity. Only the identity fixes a codimension-1 face.

(4) The automorphism group of U is transitive on codimension-3 faces.

Examples of such cell complexes include

- the n-simplex
- the n-cube
- the icosahedron
- the dodecahedron
- the n-dimensional octahedron
- the 24-cell
- the 120-cell
- the 600-cell

A U_1-complex is a cell complex whose top cells are isomorphic to U_1. These isomorphisms are the structure maps. Suppose that M is a U_1-complex whose topological realization is a manifold. For instance, U is such a U_1-complex. Given a map on one codimension-1 face, (3) implies that there is exactly one extension across any codimension-2 face. Consequently, up to isomorphisms of U_1, M has at most one map to $\mathbf{U}_1$.

Define $\hat{M}$ to be M with a copy of **U** added to each codimension-1 face of M. A coloring of M is a map from M (or $\hat{M}$) to **U**. If F is a codimension-2 face of M, let ρ(x) be the number of top dimensional faces of M containing x. In case M is **U**, every codimension-3 face of **U** lies in the same number of top faces, by (4). Let this number be denoted u. We get an interesting theory when u is odd. In the list of examples, u is odd except for the 24-cell and the octahedra. The degree of a coloring is defined as usual.

Theorem 1

If u is odd, then mod 2, deg(f) is the number of odd codimension-2 faces of M mapping to a fixed codimension-3 face of **U**.

Proof:

If f is a coloring, then mod 2, deg(f) is the number of top simplices mapping to any face. We have then

$$deg(f) = u \cdot deg(f) = \sum_{F \subset T} (\text{number of faces of M mapping T}) = \sum_{f(c) = F} \rho(c)$$

where F is a codimension-3 face of **U**, T a codimension-1 face of **U**, and c is a codimension-2 face of M. □

Corollary 2

If u is odd and if the number of odd codimension-2 faces of M is less than the number of codimension-2 faces of **U**, then every coloring of M has even degree.

A $\mathbf{U}_1$ complex is called even if every codimension-2 face has even degree. If M has a map to $\mathbf{U}_1$, then M is even. This follows from the fact that the mapping reverses orientation at each codimension-1 face. If **U** and M are simply connected, then if M is even there is a map from M to $\mathbf{U}_1$.

We can define $U_1 \times U$, and the result is a U_1 complex where every codimension-1 face lies in exactly two top faces. It follows that the top homology group is $\mathbb{Z}$ (since M is oriented). If M has a map e to U_1, and f is a coloring of M, then we get a map $e \times f: M \to U_1 \times U$. The degree of the projection $U_1 \times U$ to U divides the degree of f. This degree is seen to be the size of $\mathrm{Aut}(U_1)$.

Theorem 3

$|\mathrm{Aut}(U_1)|$ divides the degree of every coloring of a globally even U_1-structure on a closed manifold.

It is easy to see that if u is odd, then the boundary of the set of non-singular codimension-2 faces is the set of odd codimension-2 faces of M. If D is a subcomplex of M, and f is a coloring of M such that $f(\partial D) = F$, where Fis a codimension-2 face of U, we can construct a new coloring g as follows: let g=f on D, and $g=\sigma f$ on the complement of D, where σ is from (3). We say that f and g are Kempe equivalent, and extend this to an equivalence relation. As before, we see

Theorem 4

If u is odd, M and U are simple connected, then $B_U(M)$ is connected.

We can define homotopy and cobordism, but the problems are hard. Here is one simple result:

Proposition 5

There is no triangulation of the n-sphere such that all odd codimension-2 simplices are contained in one top simplex, unless there are no odd simplices.

Proof:

The complement is simply connected, and so has an n+1 coloring. This gives a coloring of the vertices of the exceptional simplex. If any codimension-2 face of this simplex is odd, following the coloring around gives the contradiction that two vertices are colored alike. □

Some comments on icosahedral colorings: A simple calculation shows that every map from the icosahedron I to itself is an automorphism. It follows that $B_I(I) \cong I$. The analog of the four color theorem fails, for there is no map from Δ^3 to I. If M has a 3-coloring, then $B_I(M)$ is non-empty.

Notation

$\hat{}$	hat	1.4
$\times$	product	1.2
$\coprod$	disjoint union	1.1
#	cartesian product	1.2
$\circ$	composition of graphs	1.5
$*$	join	1.4
$X \int B(Y)$	wreath product of X and B(Y)	1.5
Δ^n	n-simplex	1.1
I^n	n-cube	7.2
$\rho(\sigma,X)$	degree of the simplex σ in X	5.2
V(X)	set of vertices of graph X	1.4
E(X)	edges of graph X	1.4
B	coloring functor	1.2
$B_{\mathbf{Z}}$	general coloring functor	7.1
$B^2(X)$	B(B(X))	1.2
Φ	natural map from X to B^2(X)	1.2
L(X)	line graph of X	4.1
K_n	complete graph on n vertices	4.1
$K_{n,n}$	complete bipartite graph	4.1
$K^r{}_m$	complete r-uniform hypergraph on m vertices	4.3
$K^r{}_{n,n,\ldots n}$	complete r-partite hypergraph	4.3
P_n	cycle with n vertices	1.3
Sym(n)	symmetric group on n letters	1.3

S_n	**simplicial complex associated to Sym(n)**	**1.3**
D_n	**the derangement complex**	**2.3**
kX	**k disjoint copies of X**	**1.2**
$X^{\#n}$	**X#X...#X (n factors)**	**2.1**
X^{*n}	**X * X $\cdots$ * X (n factors)**	**1.5**

BIBLIOGRAPHY

Alspach, B., "A 1-factorization of the line graph of complete graphs," *J. Comb. Theory*, vol. 6, pp. 441-445, 1982.

Baranyai, Zs., "On the Factorization of the Complete Uniform Hypergraph," *Colloq Math Soc. Janos Bolyai*, vol. 10, pp. 91-108, North Holland, 1973.

Biggs, N., "Pictures," in *Inst. of Math and Its Application*, pp. 1-17, Essex, England, 1972.

Bose, R. C., "Hermitian Varieties in a finite projective space PG(N,q squared)," *Canad. J. Math*, vol. 18, pp. 1161-1182, 1966.

Bruch, R. H., "Quadratic Extensions of Cyclic Planes," *Proceedings Symposium in Appl. Math.*, vol. 10, pp. 15-44, 1960.

Cameron, P.J. and J.H. van Lint, *Graphs, Codes, and Designs,* Cambridge University Press, Cambridge, 1980.

Castagna, F. and G. Prins, "Every Generalized Petersen Graph has a Tait Coloring," *Pacific J. of Math*, vol. 40, pp. 53-58, 1972.

Cofman, Judita, "Baer Subplanes in finite Projective and Affine Planes," *Can. J. Math.*, vol. 24, pp. 90-97, 1972.

Coxeter, H.S.M., *Regular Polytopes,* Methuen and Co., London, 1948.

Coxeter, H. S. M., "My Graph," *Proc. Lon. Math. Soc.*, vol. 3, pp. 117-136, 1983.

Denniston,, "Some Packings of Projective Spaces," *Lincei-Rend Sc. Fis Mat e Nat*, vol. 52, pp. 36-40, 1972.

Fisk, Steve, "Geometric Coloring Theory," *Adv. in Math*, vol. 24, pp. 298-339, 1978.

Fisk, Steve, "Variations on Coloring, Surfaces, and Higher Dimensional Manifolds," *Adv. in Math.*, vol. 25, pp. 226-266, 1979.

Fisk, Steve, "Cobordism and Functoriality of Colorings," *Adv. in Math.*, vol. 37, pp. 177-211, 1980.

Fisk, Steve, Daniel Kleitman, and D. Abbw-Jackson, "Helly -type Theorems about sets," *Discrete Math*, vol. 32, pp. 19-25, 1980.

Fisk, Steve, "Automorphisms of Graphs," *Congressus Numerantium*, vol. 38, pp. 139-144, 1982.

Fisk, Steve, "Latin squares of Automorphisms," *Congressus Numerantium*, vol. 39, 1983.

Godsil, C. D., "The Automorphism Groups of some Cubic Cayley Graphs," *Europ. J. Comb.*, vol. 4, pp. 25-32, 1983.

Hartshorn, Robin, *Foundations of Projective Geometry,* Benjamin, New York, 1967.

Himelwright, P. E. and J. E. Williamson, "On the 1-factorability and Edge colorability of Cartesian Products of Graphs," *Elem. Math*, vol. 29, pp. 66-68, 1974.

Hirschfeld, J. P., in *Projective Geometry over Finite Fields*, Clarendon Press, Oxford, 1979.

Jaeger, F., "Sur l'indice chromatique du graph representatif des aretes d'un graph regular," *Disc. Math.*, vol. 9, pp. 161-172, 1974.

Kotzig, Anton, "Problems and recent results on 1-factorizations of Cartesian products of graphs.," *Congressus Numerantium*, vol. 21, pp. 457-460.

Lovasz, L. and D. Greenwell, "Applications of Product Coloring," *Acta Math Acad. Sci. Hung.*, vol. 25, pp. 335-340, 1974.

Parsons, T., "Graphs from Projective Planes," *Aequationes Math*, vol. 14, pp. 167-189, 1976.

Rosa, Alexander, "Systems," *J. Comb. Th. (A)*, vol. 18, pp. 305-312, 1975.

Ryser, H., in *Combinatorial Mathematics*, J. Wiley and Sons, New York, 1963.

Sabidussi, G., "Graph Multiplication," *Math. Zeit.*, vol. 72, pp. 446-457, 1960.

Sotteau, Dominique, *Decompositions de Graphes et Hypergraphs*, 1980.

Teh, Hoon Heng, "Fundamentals of Group Graphs," *J. Nanyang Univ*, vol. 3, pp. 321-344, 1969.

Thas, Joseph A. and J. Chris Fisher, "Flocks in PG(3,q)," *Math. Zeit.*, vol. 169, pp. 1-11, 1979.

Vizing, V. G., "On an estimate of the chromatic class of a p-graph," *Diskret Analiz*, vol. 3, pp. 25-30, 1964.

Wallis, W. D., *One-factorizations of Wreath Products*, pp. 337-345.